The Packet Radio Operator's Manual

114   ICOM HT conn.

146   C20 ,C21

The Packet Radio Operator's Manual

By Glynn ''Buck'' Rogers, Sr., K4ABT

CQ Communications, Inc.

Second Printing 1995

Library of Congress Catalog Card Number 93-71169
ISBN 0-943016-04-5

Editor: Gail M. Schieber
Layout and Design: K & S Graphics and Hal Keith
Phototypographer: Pat Le Blanc
Cover photo: Larry Mulvehill, WB2ZPI

Published by CQ Communications, Inc.
76 North Broadway
Hicksville, New York 11801 USA

Printed in the United States of America.

Table of Contents

Credits, Trademarks, and Copyrights

TAPR: Tucson Amateur Packet Radio is a non-profit research and development group dedicated to advancing amateur digital communications.

AEA: The registered trademark of Advanced Electronic Applications.

ATARI: A trademark of ATARI Corp.

Apple and Apple Macintosh: Trademarks of Apple Computers Inc.

IBM and IBM PC: Trademarks of International Business Machines Corporation.

Commodore 64/128: A trademark of Commodore Inc.

DRSI: A trademark of Digital Radio Systems Inc.

Kantronics: The trademark of Kantronics, Inc., Lawrence, Kansas.

MFJ: A trademark of MFJ Enterprises, Inc.

MULTICOM.EXE: Copyright Bob Slomka.

Net/Rom: Copyright Software 2000 Inc.

PacComm: A trademark of PacComm Packet Radio Systems, Inc.

ROSE: Copyright RATS Open Systems Environment, and written by Tom Moulton, W2VY.

TRS-80 and TRS 80 CoCo: Trademarks of Tandy/Radio Shack Corp.

Beyond The Meaning
Of The TNC

This manual is not intended as a beginner's guide or as a packet primer. It is a manual for the packet radio operator who has weathered the grind of studying the "how to's" and the basics.

You've already become accustomed to the catchword, or acronym, **TNC** (Terminal Node Controller). It was the first packet radio buzz word you learned, and it helped you define how much (or how little) you knew about the packet station. If by chance you missed that definition at the beginning of some of my earlier books about packet radio, the TNC is a device with a microprocessor as its brain. This microprocessor is fed from a spoon which contains a special diet consisting of a slate of instructions called "firmware." This firmware enables the microprocessor to manage many kinds of binary or ASCII information. The "spoon" is called a **ROM**, or Read Only Memory. Many companies tailor their TNCs by using an **EPROM**, which means Erasable Programmable Read Only Memory. This latter type of "spoon" allows the vendor to periodically update the features and command structure of the packet controller.

The TNC interfaces the terminal or computer via the communications port. On another port of the packet controller is the radio or transceiver. It is what happens to the data inside that gives the TNC all that

The trademark of K4ABT, and the author's favorite image transferred via packet. Note that the photo is error-free. Although the photograph is shown in black and white here, the picture was received and displayed in 256 vivid colors.

amazing power. To be very brief, it gives a personality to the otherwise senseless data.

First the data has a **FLAG**, or personality bit sequence, added to it. Next there is the **ADDRESS** field; this tells the packet where to go and how to get there. A **CONTROL** byte is then applied to the packet; it tells it how much to carry, how long to carry it, and what to bring back to prove that it has completed the mission (see figure I-1).

In the meantime, over in the corner of the TNC is the accountant waiting for the return acknowledgment, or **ACK**. This accountant has a special helper called the **FCS**, or Frame Check Sequence, to check for errors in the returned ACK. If the FCS finds a bit-count error, it will tell the accountant, and the packet will be sent again, until the correct FLAG bit is returned. This final scenario is what gives packet its error-free personality.

Before we get too far ahead, we need to fill the TNC-to-radio gap. Let's talk in terms of VHF, since this is where 85 to 90 percent of the new packeteers begin their packet hobby.

The TNC interfaces to the 2 meter rig via three lines (see figure I-2). Transmit audio (AFSK) is fed from the TNC to the radio microphone input. Receive audio is fed to the TNC audio input from one of the following sources:

1. The receive discriminator
2. The receive detector
3. The speaker output (most used source)

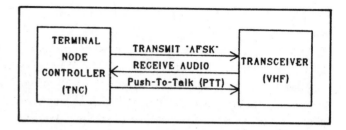

Figure I-2. Three lines are necessary for communications and control between the TNC and the transceiver. Direction of signal and control flow is indicated by the arrows.

The third line is the Push-To-Talk (PTT) line from the TNC, which provides automatic control of the transmit and receive function when a packet is sent from the TNC. This control is activated when a carriage return is issued or when the PACLen is reached. This action can occur when the TNC is in either the **CONVERSE** (CONV) or the **TRANSPARENT** (T) mode.

Protocol

The primary consideration is that your TNC should support AX.25 Level 2 Version 2 Protocol. The latest production TNCs do support this protocol. Most TNCs have a command called **AX25L2V2**. In the older TNCs this command is usually defaulted to OFF, but in the late-model TNCs you will find that it is now defaulted to ON. When you begin configuration of

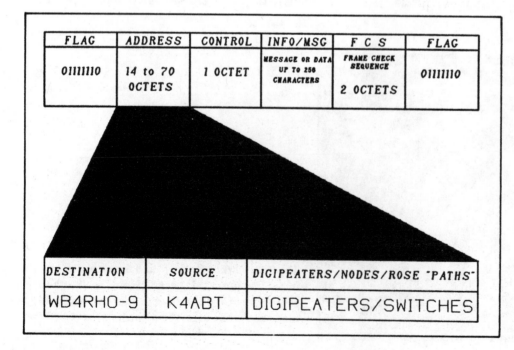

Figure I-1. The illustration above depicts how the AX.25 version 2 packet frame is formatted.

your TNC, this should be one of the first commands that is set to ON.

The second consideration relates to the kind of computer or terminal you are using. All TNCs support an RS-232 terminal interface. However, the Commodore C-64 does not support true RS-232. Many C-64 terminal programs enable the user port (the edge connector on the left rear as you are looking from the front) to support ASCII communications with TTL levels (0 and 5 volts). Most packet controllers have provisions to interface to a TTL port such as that implemented in the Commodore.

If you plan to use one of the controllers which do not support the TTL communication port, then plan for the expense of a TTL–to–RS-232 converter for the C-64.

Without going into the explanation of the RS-232 standard here, I've prepared Addendum A at the end of this book to explain the RS-232 signal format more fully.

Becoming Familiar With The "Command" Buzz Words of The TNC

So often the packeteer is confronted with a buzz word or acronym that is alien. Such is the case of the acronym used to indicate the command state of the TNC. The acronym is the simple **CMD:** which appears on the display when the TNC and terminal are set to the correct baudrate, data word length, and parity. CMD: denotes that the TNC is in the COMMAND state and is ready to receive input from the terminal.

Learn the difference between the **COMMAND** (CMD:) and **CONVERSE** (CONV) mode. In the COMMAND mode you are talking to the packet controller and anything you type is interpreted as a command by the TNC. In the CONVERSE mode you are talking to the packet world, and anything you type is converted into packets and transmitted over the air as a packet of data every time you press < enter > or carriage return (C/R). To get into the COMMAND mode type **CTRL-C** (holding down the control key while momentarily pressing the "C" key). You know you are in COMMAND mode when you see the familiar prompt **CMD:**. If you are in the COMMAND mode and you wish to go into the CONVERSE mode, type the command CONVERSE (or CONV for short). Many TNCs permit the use of the "K" key to enter the command mode. This too will take you directly to the converse mode without typing CONVERSE.

A large number of terminal programs for the PC or compatible use the **F1 and F2** keys to enter the COMMAND (F1 = CMD:) and the CONVERSE (F2 = CONV) modes, thus reducing the command key inputs to only one keystroke. Remember: When you want to change or modify a command in the TNC, you must be in the COMMAND (CMD:) mode. Reading the manual that comes with your packet controller is helpful in learning the first necessary lesson involved in setting the commands within the packet controller.

Realizing there are over 100 commands in some TNCs, we soon learn there are also just as many ways to configure the TNC. It is my responsibility here and now to inform you that in no way, shape, or form are you to try to recall from memory every command for your TNC. As a matter of fact, there are only five or six parameters that need to be set in a new TNC. With the exception of your callsign, the bulk of the commands are defaulted to the correct setting.

I will not try to "plan your lunch," so to speak. In other words, I'll not instruct you as to which command you should set, and to what you should set it. In chapter two of this manual I will assist you in the understanding and use of the TNC commands. For now, I'll acquaint you with how I set and use a few parameters in my personal TNC. It will be up to you, the manual reader, to try them for fine-tuning of your own applications. As you become familiar with your TNC, you may discover better ways in which to configure your own TNC. This allows you to refine the setup to suit your needs and transceiver requirements.

I Begin This Way

As soon as I have the terminal talking to the TNC, I set "**MYcall** K4ABT." If I'm using the PC or compatible with a split-screen terminal program, I set **ECHO OFF** unless the **DUPlex** in the terminal program is set to **FULL**.

I prefer an **ABaud** (terminal to TNC baudrate) to be set as high as the PC will handle. At best, I will not use a baudrate below the **HBaud** (TNC to radio baudrate). A rule that we try to maintain is to set the ABaud at least equal to, or greater than, the HBaud. This allows the terminal buffer(s) to keep the TNC buffers clear of incoming data.

Here is a word of caution for those who like to read the scrolling screen. If the ABaud is too high and the data flow coming in is from a BBS with a long PAC-Len, you may find it difficult to read the scrolling screen. The best way to handle this is to let the data flow into the buffer and save it to disk. Later you will be able to print it, or "page" through it in a text editor.

Many computer terminals maintain an internal handshaking that does not allow the screen to scroll too fast. If this is the case with your terminal, and if you are running an HBaud of 2400 bps, then by all means use an ABaud of 2400 minimum.

Note: The Commodore VIC and C-64 are TTL controlled at the I/O data port, and all strapping options or connections to the TNC should be made accordingly. The TRS-80 Color Computer and the PC or clones are standard RS-232 serial data streams.

Timing Is Important

There are several timing parameters which you need to set properly. Check your manual for the correct millisecond timing conversions.

TXDelay is the delay interval between key-up and start of data transmission. Normally, 300–400 milliseconds is adequate (TXD 30–40), but some 2 meter rigs take a bit longer to get up to speed after the keying line is asserted. If you seem to be having a problem being heard and all else seems okay, try increasing TXDelay to 400–500 milliseconds (TXD 40–50).

The parameters RESP (Response Time) and DWait are assigned to individual users to allow staggering of ACKs. **RESP** is the time delay between reception of a packet and transmission of an ACK, and **DWait** sets the delay between when activity is last heard on the channel and key-up. You will be requested to set values of RESP and DWait in milliseconds. Don't forget to convert the proper command value in your TNC. Normally, the DWait is defaulted to the correct setting of 16.

FRACK may be set to 4 and RETRY between 7 and 10. *Do not* set the RETRY to "0." To do so can wreak havoc if you try a connect to a station who is not on the calling frequency and you forget the connect attempt. **FRACK** sets the number of seconds *between* tries, and **RETRY** sets the number of times your TNC will try a packet before it gives up.

Chapter 1

The First Question

Do you remember the first question we raised when we began trying to understand data transfer? "Why can't we transmit data as it comes directly from the Data Terminal Equipment (DTE)?" Data in its raw form has no real handshaking or speed limit. At the same time, it has no rhyme or reason as to where it is headed. But most significant to our packet system is the lack of any error-checking headers or protocol. Here is where we arrive at a need for the packet TNC, or **Data Communications Equipment (DCE)**.

As I've already pointed out in the introduction to this book, most packet operation begins at VHF. Our VHF transceivers are designed for an audio bandpass of approximately 300 to 3000 cycles (Hz). The packet controller in turn takes the raw data from the I/O device or DTE and refines it into usable packets of information. Some of this information we never see, nor do we need to see it. Some of the refinements are the addition of the packet length (PACLen), speed (baudrate), Forward Error Checking/correction (FEC), and handshaking (RTS/DTS). This latter feature of the packet controller is a way of telling the DTE "hold the data stream until I am ready for more."

A TNC Is A Converter

The data entry system you are using to input data is a digital-based device. We cannot make our present-day transceivers accept this digital data in the pulse form as it comes from the DTE without some loss to the pulse shape—and in turn, loss of data.

Here is another reason we need the packet controller (TNC). It is here that the data pulse-train is filtered and reshaped from digital to analog signals and fed to the transceiver microphone input as an AFSK or QPSK tone(s). You can now begin to see the true function of the packet controller.

The ones and zeros contained in the data stream from the DTE are converted to tones, or analog waveforms which resemble tones, that lie within the 300 to 3000 cycle (Hz) bandpass of our transceivers. The reverse of this procedure is true for the incoming data as well. The DCE converts the analog signal from the radio into filtered, reshaped, error-checked, and displayed data.

A Quick Review

The best part of the amateur radio hobby is the least complicated. Truly, many of us are already equipped with at least two thirds of the essentials that make up a packet station. The home computer—such as an Apple, Atari, CoCo, Commodore, IBM PC, or clone—is one third of the packet station.

An FM rig, handie-talkie, or 25 watt transceiver makes up the second part of our packet station. The final part of the packet station is the heart of the system, which is called the Terminal Node Controller, or simply the TNC.

The Pakratt Model PK-232MBX is one of the early dual-port multi-mode data controllers. (Photo courtesy Advanced Electronic Applications, Inc.—AEA.)

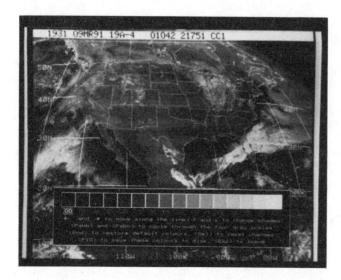

A very popular mode of communication among amateur radio hobbyists is the receiving of weather facsimile (WeFAX) photos. (Photo courtesy Advanced Electronic Applications, Inc.—AEA.)

Compared To Other Modes Packet Is Less Expensive

Compared to some other modes and facets of amateur radio, packet is not expensive. Therefore, the entry of packet into the hobby lends itself to a modular approach.

We talk about the three modules that it takes to make up a packet station. The transceiver is at the top of our list, as it seems to be the first item we obtain when we begin our entry into amateur radio. After a time we come up with some good reason to install a computer in the shack. Finally, we read about packet radio in a publication such as *CQ* magazine or a handbook that we receive at a packet forum. Here is when the third module makes its appearance in our shack.

TNCs come in various configurations. The most popular TNC is the packet-only TNC. There are others which boast integrated and/or added features such as facsimile (WeFAX), NavTex, RTTY, AMTOR, CW, and SSTV. That's correct: some TNCs offer the slow-scan television mode. These terminal node controllers go beyond the packet application(s). Thus, they are often referred to as "multi-mode" or "all mode" controllers.

All current packet controllers now have personal mailboxes incorporated in their firmware to allow messages to be left for the control operator while he or she is away. An added feature with the mailbox, or message system, is an indicator that lets the oper-

ator know when mail is in the (electronic) mailbox addressed to his/her callsign. This indicator is in the form of a flashing LED or an audible tone to get the owner's attention.

The HF Users Along The Network

Above 28 MHz the baudrate (data speed) changes rapidly. Above 2 meters the baudrates can become mind boggling. Many VHF stations are now moving to 9600 baud as a "user" baudrate, while backbone and trunk speeds are topping 100 "million instructions per second" (MIPS).

Below 28 MHz 300 baud is the maximum baudrate allowed. However, that may soon change (increase). I should point out that HF packet at 300 baud occupies a bandwidth that is closer to 500 Hz wide than the 2 kHz we continue to use.

If we were to use CW filters in some of our transceivers, we would perhaps find that more spectrum utilization can be achieved. With the 2.4 to 2.7 and in many cases 3.0 kHz single-sideband filters in our HF transceivers, we could operate as if we had a one kHz spacing or guard-band. There is one drawback, and that is the "buck-shot" effect from strong stations on a nearby packet frequency. This is one of the reasons why so many would-be HF packeteers complain of too much interference on the HF frequencies.

The answer is to locate a 500 Hz IF filter that is usually used for CW operating. Most modern-day transceivers allow these filters to be installed so that they may be switched on and off, or in and out of operation. In some Kenwood transceivers there is a filter switch dedicated to the switch selection of an **SSB NARROW** position. Be sure you have the correct type of *crystal* filter installed in the SSB NARROW slot. By using this method of spectrum conservation we

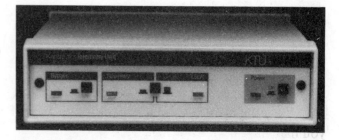

The Kantronics KTU is used in conjunction with the weather node to access weather information via packet. (Photo courtesy Kantronics Inc.)

may discover that we now have nearly a kHz of guard-band between channels.

Once you have your transceiver operating with a 500 Hz IF filter, and the one kHz spacing, you have a decent guard-band. The only proviso is that it is like the soap we hear about: "We use XYZ soap. Don't you wish everyone did?" If everyone were to use these sharp-tuned filters, HF operating would be much more enjoyable, not to mention it would reduce the number of tries, reducing the amount of congestion.

"Quick-Connect" of The Third Kind

When power is applied, packet radio TNCs are normally in the disconnected state and/or monitor mode. This allows the activity on the channel to be seen by the terminal display. In addition, the TNC will see any attempt to connect (called a "connect request") to your station when it is in the **MONITOR (MON)** mode.

Let's take a short walk through a packet connection. When station one wishes to connect to station two, a connect frame is issued from the command mode and a time-out timer is set into action. If station number two is on the air, a "connect" is established, and station number two returns an acknowledgment (**ACK**) frame. If station number two is not on the air or for some reason it doesn't respond, the timer runs out, and station number one will display

"Retry count exceeded!!!
***DISCONNECTED**

After a link connection is established, the TNCs will enter the data transfer or the **CONVERSE (CONV)** mode. In this state the two stations can exchange data or other digital information in any number of ways, from the good old rag-chew session of the CONVERSE mode to the transfer of high-resolution, color pictures that appear on the screen as they are received, but in the transparent mode.

Where Do We Begin Operating Packet?

Now that we have given you an entree into this mode of communications, we are ready to show you how very easy and interesting packet radio really is. You will soon discover that packet is not the hurdle that you may have been led to believe it is.

Nearly 85% of all packet radio communications take place on the VHF bands, with approximately 75% of the operation on the 2 meter band alone. The con-

AEA-FAX is an IBM-compatible hardware and software package for the demodulation and display of HF WeFAX information. (Photo courtesy Advanced Electronic Applications, Inc.—AEA.)

ventional 2 meter FM transceiver or handie-talkie represents one third of the required packet station. In addition to the FM transceiver, the packet station consists of a computer or video terminal and a TNC. The task of the TNC is to send and receive information to and from the computer or terminal so the incoming and outgoing data can be displayed, buffered, and/or saved. We use repeaters for packet just as we do with VHF FM voice repeaters. We add a slight twist to the method used to transmit the data to a distant station. The conventional voice repeater is a duplex system that transmits (repeats) the signal on a frequency different from the frequency that is being received.

Packet radio uses a system that is fundamentally different, because only one frequency is utilized. This means less equipment, and thus a less complicated (simplex) installation.

The packet repeater is called a digital repeater, or more commonly, a **digipeater.** The digipeater is a store-and-forward device that receives a packet of data (usually 128 bytes or characters) and stores it to an internal buffer. It then "listens" to be sure the frequency is clear before it transmits (forwards) the packet to the next digipeater or the target station (see figure 1-1).

The digipeater does not require a duplexer or other distinguishing control circuits that you would find in a voice repeater installation. A digipeater consists of a transceiver and a TNC. The digipeaters and nodes (a **node** is a new "super" digi that stores routes to other digipeaters and nodes) are usually located on high buildings or mountaintops. A digipeater normally consumes an insignificant amount of power and can

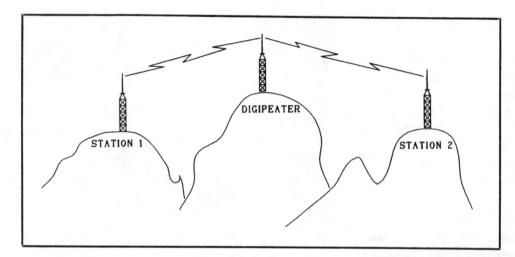

Figure 1-1. The digital repeater (digipeater) is a store-and-forward device that receives packets (normally 128 characters per frame) into an internal buffer. Once the buffer is filled, or the PACLen is reached, the digipeater listens to make sure the frequency is clear. If there is nothing heard, the digipeater then forwards the data to the next digipeater, or the destination station.

easily be mounted inside a ventilated box on a tower and sometimes at the base of the antenna.

By using more than one switch, node, digipeater, and/or gateway in my packet connect paths, I am able to cover great distances with very low power. As a matter of interest, in my briefcase I have a small lap-top portable computer, TNC, and 5 watt handie-talkie which I use to communicate with friends at the top of the world—over 8000 miles away.

When a digipeater is used to establish a contact (connect) to a distant station, the callsigns of the digipeater will appear in the packet, along with the callsigns of the sending and receiving stations. Once the packet is received by the digipeater, it is analyzed, and if the packet is error-free, it is then forwarded to the callsign of the target station or the next digipeater.

Chapter 2

Building A Packet Station Is Easy!

Remember when you beat the UPS person to the door, accepted the parcel, and turned to the task of ripping and tearing open the box to see this long-awaited device called a TNC? You removed the device from the box. With a gleam in your eye you held it, and in some distant memory you recalled a childhood moment that was more real than fantasy. At that point you entered another world of communications that has become more fun than can be described.

For a few moments let's return to that day and savor those moments just in case you have forgotten how sweet it was.

The packet industry has come to you in the form of a box full of firmware (that's the name given to lots of software captured inside an IC called an Erasable Programmable Read Only Memory, or just plain EPROM). Before you get too deeply involved in the text of the packet controller manual, it is first necessary to look at some of the small but important items you may need to complete your packet station.

Most packet controllers come with the cables and connectors for the controller end of the interface cables. The opposite end of the cables now becomes important to the project.

The packet controller manufacturer has no idea what kind of radio or computer terminal you might choose or already have chosen. There's no simple means that will provide a way to supply the correct connector for your RS-232 port or the microphone connector for your transceiver.

Checking the list further, you observe a need for some type of terminal program. Most terminal programs for the telephone modem will work, but there are many custom-written programs for packet which work much better. You may already have a favorite software package that you wish to use instead.

Patience Is The Watchword

The first thing you do with a project of this nature is define the task. You soon discover that the project flows together with much greater ease than you had first imagined.

It was once a chore to find the special connectors for some of the imported transceivers. This is no longer the case. A trip to the local supply house will prove refreshing when you discover the many types of connectors for both the computer and the transceivers. Many electronic parts centers now stock the popular 8-pin connectors used by Alinco, ICOM, Kenwood, and other transceiver manufacturers (see figure 2-1; many other radio and TNC connections are illustrated in the "Transceiver-to-TNC Connections" section of this book).

If your radio does not provide for receive audio at the MIC connector, then you should use the (3.5 mm jack on most makes) external speaker jack for the TNC receive audio. Observe the wiring colors in the

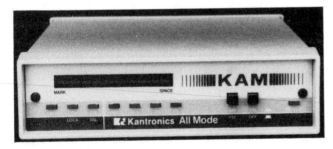

The KAM is an all-mode communicator with HF to VHF gateway capabilities. (Photo courtesy Kantronics Inc.)

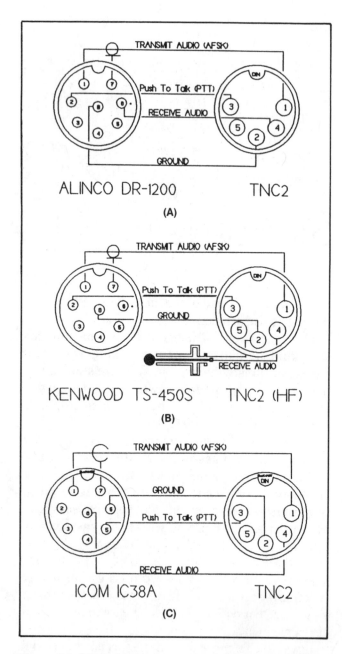

Figure 2-1. Most electronic parts houses stock the popular 8-pin microphone connector.

cable from the TNC to the radio. Be sure the grounds are indeed grounds, and the signal shields are connected as required. The reason I mention this is because some radios use a floating shield for the audio input, separate from the system, PTT ground.

RS-232 or TTL ?

Now that the transceiver and TNC are connected, let's turn our attention to the TNC-to-computer interface. Most TNCs provide the means to connect to, and op-

erate with, both TTL and RS-232 communications ports. There are a few TNCs that require an additional TTL to RS-232 converter. Such a signal converter will be needed with the Commodore 64 and some Atari models when used with some of the earlier type TNCs. Most TAPR II and clones support both kinds of interface hardware.

Let's assume you plan to interface your TNC to the standard RS-232 port. All TNCs support the RS-232 standard in one form or the other. I have illustrated the EIA RS-232 standard functions in figure 2-2.

Typically, we would quickly use a ready-made cable, but this time we need to stay within the rule and build our cable with the one furnished, or use the one described in the packet controller manual. If it is the standard DB-25 connector and RS-232 standard, then use pins 2, 3, 4, 5, and 7. These wires will go pin to pin DB-25 to DB-25 connector. Observe the gender of the computer serial or comport connector, and above all *don't* make the mistake of plugging your TNC cable into the PRINTER port of your PC. It is wise also to check the kind of connector used on the TNC. Some dual-port TNCs use connectors other than the standard DB-25. To add some fun to our plight, some computers have adopted connectors of their own design, and you will find that these connectors are *not* easy to find (see figure 2-3). Some of the new computers are using the 9-pin (DB-9) connector, since the manufacturers have found that only five signals are necessary in most communications applications (see figure 2-4; for other computer-to-TNC interface configurations, see the section "Computer Communications (Serial) COMPorts To TNCs" at the end of this book.

Recheck the wiring of both cables. It might even be worthwhile to check the manufacturer-supplied cable wiring. After you complete the checklist, connect the cabling to the respective components and turn on the TNC. If all works in the manner the manual says it should, then "boot" your terminal program, and set the parameters to the default parameters of the TNC, if possible. Many TNCs have default parameters set as follows: Baud (ABaud) 1200 (9600 bps is preferred); word length (databits) 8; stopbit(s) 1; parity NONE (if word length is 7, then use EVEN).

You should now place your computer into the terminal or communications mode. Next turn the TNC off for a moment and turn it back on. Some characters should appear on the screen; these can vary from garbage to plain text. Let's examine what you may see. If the first thing you see is a sign-on message and the

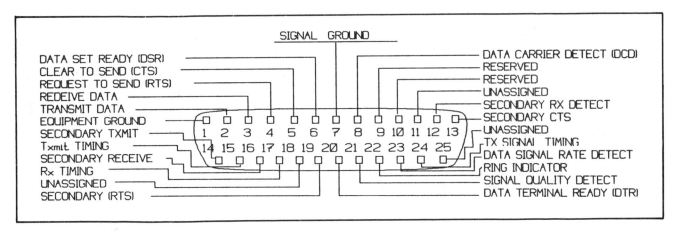

Figure 2-2. This illustration lists the signal titles as applied to the RS-232 standard. The type of connector has no relationship to the RS-232 signals.

prompt CMD:, you have really gone beyond the call of duty. If you are unable to get the sign-on message at first, don't be alarmed. Check the TNC manual for the section dedicated to the TNC setup. As soon as you have the CMD: prompt on the screen, you should enter your callsign—e.g., at the CMD: prompt type **MYCALL [your call]**. If you are seeing double letters displayed to the screen, set **ECHO OFF**. If you want to have a faster screen display, then the next step is to set the ABaud (TNC to computer speed) to a higher baudrate. To make the change, set the DIP switches on the TNC to the appropriate setting, or if you are using a TNC that has the AUTOBAUD feature, you can let it set itself to your terminal-to-TNC ABaud rate, and then turn the **AUTOBAUD** command off.

A good terminal program enables the user to set the terminal parameters on the fly. If you have a terminal program that allows you to save the current configuration, do so now. In addition, recycle the system again by turning the TNC off for a few seconds and then turning it on again.

Setting The Transmit Audio

One of the big problems we have found is that many users feed a signal that is too high from their TNC to their transceiver microphone input, and the result is over-deviation and distortion, and in general the system responds in an erratic manner. It is wise to remember that most 2 meter transceivers provide some form of limiting in the mic input audio path. If your input audio level is too high, you could be driving this limiter section with too much audio.

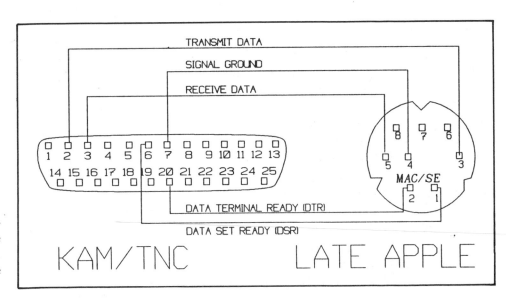

Figure 2-3. Some serial comport connectors may be more difficult to find than others. Note the connector for the MAC/SE shown here.

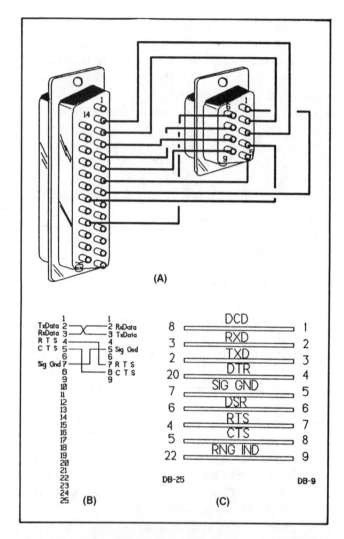

Figure 2-4. (A) Many late-model computers use the DB-9 connector as the serial comport connector. At (B) the five most-used RS-232 signals are illustrated. At (C) all nine signals are indicated in a DB-9 to DB-25 adapter.

Diodes are generally used in the limiter circuits. Thus, the end result is rectification of the audio. This results in a DC component which introduces an equal amount of phase distortion within the audio path. Distortion can cause clipping that manifests itself enough to prevent the system from communicating at all. Audio distortion is one component that packet radio will not correct, nor will it tolerate it.

Modern VHF and UHF rigs use low Z (normally 600 ohms) microphones such as the condenser or electret microphone. The microphone inputs require an audio level somewhere in the range of 10 to 20 millivolts peak-to-peak.

If you are fortunate enough to have access to an FM deviation meter, set the transmit audio level of the transceiver and TNC to 3.5 kHz ± .5 kHz. DO NOT EXCEED 4 kHz. If an FM deviation meter is not available, you may wish to use the following technique temporarily.

Most TNCs have a **CALIBRATE** or similar command that allows you to key the rig and send a packet tone. Listen to your transmitted signal with another rig, and set the level until the perceived volume stops increasing. (Be sure you adjust the mic gain control and not the deviation control.) The mic gain control is before the limiter and the deviation control. Back off the input level (level of audio from the TNC to the 2 meter radio) until you hear the volume decrease. Again, I caution you against too much audio. You will soon discover that packet radio will work much better with less than 4 kHz than it will with over 4 kHz deviation. Too much audio drive may create more distortion, and we've already covered what happens when there is distortion in the packet signal path. Turn the audio down to a point where it sounds clean when it is received on another receiver. This will be close enough for practical and temporary operation, but you should set the transmit audio with the proper test equipment as soon as possible. In short, the audio should sound clean with no distortion or harshness.

TAPR TNC-2 and clones have an adjustable pot on the printed circuit board that enables fine adjustment of the transmit audio. Other TNCs have fixed output levels that are selected with varied jumper positioning. These jumper positions can sometimes render a level that is close to that needed to make contact. On occasion one of these jumper settings may prove to be too much, and other problems can develop. You may have to set the mic level control inside the rig as described earlier.

It is advisable to check the modulation of your FM transceiver with a deviation meter to be sure there is no adjacent frequency interference.

Taking The Mystery Out Of The Packet Commands

Some of the command definitions covered herein may not apply to all TNCs, so it goes without saying that it is *not* my intention to break out each TNC and its associated commands. For the most part, however, these commands are universal except for the feature commands. We will cover each definition in the order in which it appears under the designated function group heading. We will cover only those commands directly associated with packet. This is not a command table for other digital modes.

For now let's consider this as a simple approach to the packet commands. Here is the command syntax divided into seven function groups:

Asynchronous Commands
Special Character Commands
Identification Commands
Link Control Commands
Monitor Commands
Timing Commands
Feature Commands

The related commands syntax will be covered under each of the headings, along with the function of that command.

Asynchronous Commands

8BITCONV: Useful when transferring files or sending control codes that use binary letters and numbers. When this command is set to ON, bit seven is allowed to be transmitted in the CONVERSE mode. The command is useful when transmitting control or non-ASCII characters. All eight bits are transmitted from the terminal to the distant station as if the packet controller were in the transparent mode. If long binary files are to be transmitted to a distant station, the TRANSPARENT mode will be more forgiving with file transfers. Terminal parity can be set to none while using the transparent mode.

ABaud: This is the common acronym derived from the AUTOBAUD routine. The ABaud is used to establish the terminal-to-TNC baudrate. The AUTO-BAUD begins the search at 300 bps. In some TNCs it is activated by entering an asterisk (*) during the AUTOBAUD search routine. The search will display:

PRESS * TO SET BAUD RATE

When this message is displayed, you have about 2 seconds to depress the asterisk (*). Once you have the baud rate set and your callsign entered, at the CMD: prompt you should type:

AB <space> **(baudrate)** <enter>

The TAPR TNC-2 and clones use external DIP switches to set the terminal-to-TNC baudrate. **Be sure the TNC is OFF when changing any of these DIP switch settings.** The Hbaud rate represents the "on air" speed (see Hbaud). With the data terminal speeds increasing, it is best to choose a terminal program that will support speeds of 1200 bps and above. The main reason for doing so is because you may find yourself trying to operate 2400 bps HBaud over the air, yet the terminal display is falling behind because the terminal will not display the incoming information as fast as it is being received.

AUTOLF: Sends linefeed to terminal after carriage return. When this command(s) is ON, the TNC will automatically add a linefeed after the carriage return is pressed, or if a carriage return is present in the file being transferred.

BKondel: Defines the method of character deleting. When BKondel is OFF, the back slash (\ ) will echo to the display or screen when a character is deleted. When ON, a backspace-space-backspace is echoed to the display. Many operators leave this command ON.

ECHO: Allows viewing local keyboard entries. If local keyboard characters are not appearing on the screen, set this command to ON. If double characters appear each time a key is pressed, set the ECHO command to OFF. ECHO is most often set to OFF when the terminal program has DUPLEX set to HALF, and ON when DUPLEX is set to FULL.

SCReenl: The numerical value entered with this command will set the format of the screen width from 0 to 255 letters or characters.

Many terminals display in an 80-column format; thus the reason for the default of 80. If your terminal display has fewer than 80 characters, you can format the screen length to match the screen width. If your terminal display is automatically formatted or given a linefeed and carriage return, simply set the SCReenl to zero (0).

FLOW: Prevents incoming (received) packets from interfering with those packets being typed or entered.

Note: When using split-screen terminal software, set FLOW OFF. If this command is ON, the local keyboard entries will not be affected by incoming packets. Once a keyboard entry is made, the terminal will stop any display of incoming messages or packets. When text is entered, the terminal will allow the incoming packet(s) to be displayed.

PARITY: Sets the data parity for terminal to computer data flow. The controller will only send serial output with eight data bits and one stop bit. By setting the parity, you are defining the disposition of the eighth bit.

TRFLOW: When ON, TRFLOW provides software flow control in the TRANSPARENT mode. The settings of START and STOP determine the type of flow control.

TXFLOW: Enables hardware flow control when in the TRANSPARENT mode. When ON, the XFLOW determines the flow control that is used during the transparent mode. If FLOW is OFF, software flow control is not used. The XON and XOFF characters are used when XFLOW is ON.

XFLOW: Enables software flow control for CONVERSE mode, when ON. When XFLOW is ON, the terminal should respond to the flow control characters that are entered as XON and XOFF. When the XFLOW is OFF, the controller will respond to hardware flow control only—e.g., CTS and RTS.

Special Character Commands

CANLINE: As defaulted, it is **CTRL-X**, and with this action the current line being typed in will be canceled. This command character can be changed by entering a new command character following the CANLINE command. This command is not often changed, and is usually left as defaulted.

COMMAND: CTRL-C, which is used to return to the command mode from the CONVERSE or TRANSPARENT mode. Since all references are made to the CONTROL C, it is good to leave the default character as is.

CANPAC: The **CTRL-Y** cancels a terminal output from the buffer. This character function is executed from the command mode.

SENDPAC: Carriage return will cause packet to be sent. The default character should be used, since HEX $0D is the carriage return. This allows a packet to be formed and sent each time a carriage return is pressed in the CONVERSE mode.

START: CTRL-Q restarts printing after CTRL-S. This control character will restart display or printing of incoming packets after they are halted by a CTRL-S. CTRL-S is the stop character.

STOP: CTRL-S will stop output to the screen or printer. Sometimes used with software handshaking, this command is most often used to hold the packets in the TNC buffer while the terminal or computer is off-line doing other tasks.

The buffer of the terminal or computer should be open when the start character (usually CTRL-Q) is invoked following the return of the computer to

packet operation; otherwise all data will be scrolled off the screen and lost.

STREAMSW: Selects the character to be used in switching between streams or ports of the controller. The selected character can be passed in the CONVERSE mode, will be ignored in the TRANSPARENT mode, and will continue through the controller as data. If you wish to change streams while in the TRANSPARENT mode, you must return to the COMMAND mode to do so.

XOFF: Will stop data flow from the controller. This command is used to stop packets from the controller to the terminal or computer. The start character will resume data flow.

XON: CTRL-Q restarts terminal data flow. When the CTRL-Q is enabled, the controller to terminal data flow will resume. The computer or terminal buffer should be open so no data will be lost when the command is invoked.

Identification Commands

BEACON: Beacon timing intervals after n*10 seconds.

Options **EVERY** and **AFTER**. A value of zero will turn the beacon OFF (not a bad idea, with channel congestion). Each increment of one is equal to 10 seconds in most controllers. Because of overcrowding, beacons are beginning to be frowned upon, except when used to make important announcements. If the option EVERY is used (Beacon EVERY), then a beacon will be sent n*10 seconds. If the option AFTER is used (Beacon AFTER), the beacon will be sent *once* after the specified interval of no link or digipeat activity.

MYCALL: To install your callsign into the controller for station identity. The callsign is the first input after you establish communications with the controller. This gives the controller an identity, and establishes a way for other stations to connect to your station. An **SSID** from 0 to 15 can be added to the callsign to distinguish between controllers you may have on other bands or locations.

MYALIAS: Alias callsign for digipeating. This command is usually used for a shortened callsign for nodes and digipeaters.

In most applications the MYALIAS will reflect the local airport identifier. The digi call for Atlanta, Georgia is "ATL." This is the airport ID for the Atlanta airport. The MYALIAS for my local digi is "ABT1," and reflects a part of my callsign. The alias is either selected to make the digi call easy to remember or for the geographic location.

HID: Issues an ID every 9.5 minutes when digipeating. When ON, an identification packet is sent every 9.5 minutes if the station is in use as a digipeater. If the HID is OFF, then the IDs will not be sent, regardless of use.

Link Control Commands

AX25L2V2: Provides compatibility between the older version of AX.25 and the new level 2 of the AX.25 protocol. When ON, the TNC will automatically adapt to the version being used by a connecting station. Most packet controllers allow digipeating using AX.25, level 2 and version 2.

CONOK: Defaulted ON, the TNC will automatically acknowledge a connect request. If the CONOK is turned OFF, connect requests from other stations will not be acknowledged and a DM (disconnect message/busy) will be sent to the requesting station. A message "connect request from XXXXX" will be output to the terminal.

CONMODE: Determines the mode your controller will be in, following a connect. It selects the mode of the TNC if the NOMODE command is OFF. When ON, the controller will automatically enter the CONVERSE or TRANSPARENT mode. This will depend on the setting of the command, either CONVERSE or TRANSPARENT.

CONVERSE: Sets the packet controller into CONVERSE mode, when connected. CONVERSE has no default; it is an immediate command. By typing a **K** or **CONV** at the CMD: prompt, the controller will go to the CONVERSE mode. Once in this mode, a carriage return will force a packet to be sent.

TRANSPARENT: To enter the TRANSPARENT mode. The TRANSPARENT mode allows the sending and receiving of binary and control characters without affecting the operation of the controller. Some control characters will lock the controller if sent in the CONVERSE mode. This problem is alleviated when in the TRANSPARENT mode.

DIGIPEAT: Default ON, allows digipeating by other stations. This is part of the foundation of packet radio. The ability to store and forward packets and communicate over long distances is what packet is based upon. The DIGIPEAT command is defaulted ON, which allows your station to be used as a relay or digital repeater (digipeater).

In using several stations via which to "digipeat," we are able to send data over long paths, even at VHF and UHF frequencies. Each station listed in the digipeat or via list will be used as a "stepping stone" to the next station until the packet has made its journey.

MAXFRAME: Establishes the maximum allowed number of outstanding, unacknowledged frames. Maximum value is between 1 and 7. A frame is usually 128 characters in length.

NEWMODE: When ON, the TNC will return to COMMAND mode when it receives a disconnect. If more than one connect exists, the controller will return to the COMMAND mode only if it is on the selected stream of the disconnect.

RETRY: Retry may be set to n times; n = 0 to 15, with 0 being infinite tries. Default is 15. FRAMES are transmitted n times before a "retried out" occurs. The time between tries is determined by FRACK.

USERS: Is used to determine the number of streams that can allow a connect. This command may be given any value between 0 and 26. This value specifies the number of streams that are available for connects. If USERS is set to 3, then a connect request will connect to the lowest available channel/stream A, B, or C. If all three streams are busy, then a < DM > packet will be issued to the requesting station. A CONNECT REQUEST and callsign of the requesting station will appear on the screen of the terminal.

Monitor Commands

HEADERLN: To insert a carriage return between the header line and the text of monitored packets. This command causes the text and header line to be separated.

MONITOR: When ON, all packet data and addresses are displayed. Callsigns are separated from the text by the ">" (greater than) symbol.

MALL: As defaulted, both connected and unconnected packets are monitored. The MALL command determines the category of monitored packets. This is the setting used if you wish to communicate as a group. The MALL command is also handy for running some types of diagnostics.

MCON: With MCON OFF, the controller will monitor only connected stations in the connected state. To monitor packets from stations other than the station to which you are connected, set MCON ON. This command is active when connected.

MCOM: When ON, this command will allow monitoring of all frame control sequences and numbering. It is active when the MONITOR command is ON. This command will allow monitoring of all frame control sequences and numbering such as frames (I), unnumbered information frames (UI), disconnects (D), unnumbered acknowledge (UA), disconnect messages (DM), and connect requests (C).

MRPT: Displays the calls from the origination and destination station in the monitor mode. When ON, all paths to and from originating or destination stations will be displayed. When this command is OFF, it will display only the calls from the origination and destination stations. The MONITOR command should be ON.

Timing Commands

AXDelay: The period the controller will wait in addition to the time set by the TXDelay. This command is time in addition to the TXDelay setting. AXDelay is used in conjunction with voice repeaters which are used to pass packet data. It is most often used when slow relay switching of voice repeaters is employed for use with packet.

AXHANG: Specifies the necessary "hang time" of the voice repeater used in packet service. If the repeater squelch tail is long, it is not necessary to wait AXDelay time after keying the transmitter if the transmitter is still transmitting.

CHECK: To set the period before checking for connection or time-out. Where n = 0, n may be set to any value between 0 and 30, with each increment specifying 10 second intervals. When n is set to a number larger than zero, a periodic "check" packet will be sent to the connected station to ensure connectivity. If there is no response, a time-out and disconnect will occur.

CMDTIME: Used to set the "transparent" escape time interval. CMDTIME may be configured to any value from 1 to 15, with each increment being 1 second. The escape characters must be entered three times within CMDTIME of each other to enter the COMMAND mode from the TRANSPARENT mode. A guard-time of CMDTIME is added after the last escape character entry is made. Any entry during this latter CMDTIME will abort the escape to COMMAND mode.

DWait: Used to set the period after a packet is heard on the channel before keying the transmitter. Where n = 0 to 255 in 10 millisecond intervals, DWait can be set to wait n*10 ms after hearing the last packet activity on the channel.

FRACK: The period to wait for ACK, after transmitting. Where n = 1 to 15 in 1 second intervals. The controller will wait FRACK seconds before transmitting again. If the number of tries is exceeded as specified in RETRY, the operation is aborted. Abbreviation of "FRame ACKnowledge."

HBaud: Establishes the actual baudrate setting for the on-the-air, station-to-station communications. This baudrate has no relation to the terminal-to-TNC baudrate. The Hbaud is selective, since we are restricted to 300 baud on frequencies below 28 MHz.

RESPtime: Sets the delay period in 100 ms intervals before keying the transmitter to send an acknowledgment. The number specified for RESPtime may be set between 0 and 255, with each increment representing 100 milliseconds. This setting then becomes the "minimum" delay period before an acknowledge packet is sent. This command may run concurrent with DWait and other random delays which are in effect.

TXDelay: Sets the period between actual transmitter turn (PTT activate) and the beginning of data flow. The TXDelay command tells your controller how long to wait (in n*10 milliseconds) after keying the transmitter, before it begins sending valid data. Following are some TXDelay settings for various types of packet stations and radios.

TXD 11—Optimum for 9600 baud. Maximum TXDelay at 9600 baud should not exceed 21, or 210 milliseconds!

TXD 30—Crystal-controlled radios with diode switching.

TXD 35—Synthesized radios.

TXD 40—Radios with relay.

TXD 45—Radios with external power amplifier.

Feature Commands

CTEXT: Used to enter connect text. To activate, set CMSG to ON, or PBBS. Connect text may be entered as any combination of letters and numbers. The connect text can be used to give directions, announcements, or instructions—e.g., Buck is not here at the moment, you may leave a message in the K4ABT-10 mailbox.

MHeard: Used to view a list of the 15 or more most recent stations that were heard by your station. The command is issued by you, from the COMMAND mode, and displayed on your terminal.

In some TNCs there is a related command called the **JHeard**, or "Just Heard" feature. If the JHeard command is associated with the mailbox, or PBBS, this feature may have an expanded command call that includes a "time-stamp" addition. In addition to the "time-stamp," there may also be a bonus to the JHeard feature called the **Jheard Long** (J L). If the J < space > L is executed while connected to a mailbox/PBBS bearing this feature, you will receive a listing of stations "Just" heard.

Summary

Space does not permit expanding on TNC commands that relate to manufacturer-specific features. Therefore, I've reviewed only the most-used commands. As you become more proficient with your TNC, you may no longer need to study these commands. However, for those times when you need the meaning or acronym for one of the commands, keep this information nearby as a ready reference.

NOTES

Chapter 3

Packet Radio Networks

As we become more acquainted with the operation of packet radio and the many useful features within the system, we also discover how packet radio covers regions and countries of the world. We find there is real meaning to the term *networking* when we begin to look into how packet radio traffic is moved so rapidly through these systems. When several of these systems are linked together, they are often referred to as a *network*.

Local Area Networks (LANs), trunks, and backbones are often combined to make up a network. There are two specific kinds of networks. One is made up of switches that allow inter-connecting from one switch to another. This particular kind of network uses switches called Net/Rom™, or TheNet nodes. The node network is an older topology that we find in use in some parts of the country. Many states in the United States have begun replacing nodes with the latest and more streamlined networking concept called the Rats Open Systems Environment©, or **ROSE**

switch. The ROSE offers true virtual circuit connects, and fulfills the need for a keyboard-to-keyboard communications link away from the high-traffic, high-throughput, network nodes.

The ROSE network topology works much like the telephone network that interconnects all parts of the world. When completed, this networking concept could make it possible to connect to any part of the world, without any concern as to how it is routed, or the path used to reach from point A to point B.

Making The Right Choice

The node concept allows for good traffic-forwarding capabilities, while the ROSE switch technology provides for easy keyboard-to-keyboard connects. In short, there is a place for both these networking formats. When planning a network, or a LAN, I recommend giving some thought to how your system will be used and which need is to be fulfilled. The ROSE

This AEA Model PK-88 TNC is capable of being converted to a ROSE switch. (Photo courtesy Advanced Electronic Applications, Inc. —AEA.)

Port switching is manual with the DSP-1232 multi-mode data controller. (Photo courtesy Advanced Electronic Applications, Inc.—AEA.)

network can be implemented if there is a need for individual keyboard-to-keyboard communications, while the node network lends itself to trunking or moving heavy traffic and large files across country.

Later in this book we will describe how these nodes and switches are actually constructed. For now, however, we will confine our discussion to how to use these two networking methods. Here then is how these two networking methods are applied to everyday use.

The ROSE Network: What Is A ROSE Packet Switch?

The ROSE X.25 Packet Switch is an advanced replacement for the common digipeater or other node-switching EPROM. The ROSE switch represents the state-of-the-art in packet networking technology using international standard protocols. It is the first amateur packet networking program that uses the International Standard protocol for packet networks based on the CCITT X.25. (CCITT is Consultative Committee International Telegraph and Telephone, an international organization that sets worldwide communications standards.) The program is burned into a 27C256 EPROM and installed in place of the system EPROM (not the state EPROM) found in a TNC-2 (or clone) packet controller. The ROSE X.25 Packet Switch runs in the TAPR TNC-2 or clone.

ROSE X.25 Packet Switch offers the following features:

• Hop-by-Hop Acknowledgments Between Switches—Provides reliability and high throughput.

• On-line Information—Information/Help SYSOP text supplied.

• FCC and Foreign Postal Telephone and Telegraph accepted AX.25 Level 2 SOURCE and DESTINATION Identification—The callsigns of both the station of origination and station of termination appear at each end of the connection.

• Proper Transmitter Licensee Identification—The ROSE switch always identifies its transmissions with its own callsign, not the callsign of a user. Callsigns traverse the network without adding extra SSID, or other changes.

• The Backbone is Entirely Transparent to Users—The SYSOP can add or remove switches in the backbone, and change callsigns, bands, or frequencies without having to inform users or modify BBS forwarding files.

• True Virtual Addressing—The user needs to know only the address of the desired exit point of the network, not all the intermediate steps.

• Network Resolute Routing—The Network Manager determines best path, eliminating the need for node broadcasting of routing information to other switches.

• Dynamic Route Selection—The Network will automatically attempt alternative paths to remote switches, based on information that the network manager provides.

• Predetermined Network Paths—The network manager tells each switch which paths to use, and it will not attempt impossible links because another switch was heard during a band opening.

• Support for Emergency Operations—A switch can be added to the network to provide service from the afflicted area without modifications to the existing network.

• Security System for Remote Control—SYSOP authentication of user who requests to view or modify configuration.

• TCP/IP Protocol Support—The ROSE switch provides full support for the TCP/IP protocols (Versions 3.0 and above).

• Full Radio Support on Asynchronous Port—The asynchronous port of a TNC can be attached to a 202, or other modems which support an RS-232 interface and radio, providing a dual-port system. The second port is AX.25 using the Asynchronous Framing Technique (AFT).

• Multi-Synchronous Ports Using TNCs—Since the asynchronous port has full radio support, it too can support one or more switches via a special (commonly available) RS-232 cable.

• Network Manager Remote Configuration—All configuration is done over the air; many parameters can also be burned into the EPROM.

• Battery-Backed RAM Configuration—All routing information is retained even if power is removed. No need for manual intervention when power is restored.

The Background

A few years ago an organization was formed called the Radio Amateur Telecommunications Society (RATS). Within this organization there are young men who stand tall in this rapidly expanding network protocol. They are Tom Moulton, W2VY, and Gordon Beattie, N2DSY, along with other members of RATS. They did their networking homework well.

From this beginning the foundation for the ROSE began. From here on I'll refer to the RATS Open System Environment only as the ROSE.

For more than a year I studied and compared to other systems the scheme of the ROSE switch and the manner in which it supported packet networking. The applications of the ROSE were compared to TheNet and Net/Rom. My conclusions are more than just a few words that can be spoken or written.

I made a decision to convert many of my nodes to the ROSE because the nodes were being used in a keyboard-to-keyboard environment. This was not an easy task. The change from nodes to ROSE switches represented more than merely turning a screw or twisting a few knobs. First, to convert the network nodes to something else meant education of the LAN users as to how the new system was to be addressed and im-

plemented. Second, it meant convincing those who were opposed to it only because it represented change.

The ROSE Offers Some Embedded Features

I first wrote about the ROSE in my column in *CQ* magazine a few years ago. At that time I did not move directly to the ROSE because it lacked real security. Another reason was because the ROSE needed some fine tuning.

That was then! The ROSE switches are spreading in all directions, and they are now in almost every state in the United States, and throughout the rest of the world, providing exceptional keyboard-to-keyboard communications. Time has allowed the author to build the many needed functions and features into the code. The ROSE now has security and password access for use by the SYSOP and/or network manager.

Here is a "quick-start" user approach to making a packet connect via a ROSE network.

A ROSE connect string consists of three parts. They are:

1. The callsign of the station to which you want to connect.

2. The callsign of your local ROSE switch.

3. The address of the ROSE switch nearest the station you wish to call.

Using the above format, here is how a user in Cookeville, Tennessee would connect to the Gallatin, Tennessee mailbox:

C N4SSB-10   V WA4UCE-3,615452
     1   2   3

There are numerous features which can be installed into a ROSE switch by the SYSOP. Here are just a couple of the features that are included in the ROSE switch.

The INFO Feature: INFO is used to give an information listing of other ROSE switches within the area code where it resides. An example of how users in the Cookeville area would access the INFO feature is as follows:

C INFO V WA4UCE-3,615372

In Lafayette, Tennessee a user can get INFO as follows:

C INFO V KC4NEH-3,615411

The HEARD Feature: To use the HEARD feature, connect to HEARD. Here is an example of how I would connect to the HEARD feature in the ROSE switch at Lafayette from the ROSE switch in Gallatin:

C HEARD V N4SSB-3,615666

Immediately you receive:

STANDBY, Call Being Processed [etc.]

In a few seconds you receive:

Call Complete, You're connected to HEARD: X.25 ROSE switch [location, etc.]

This feature gives the user a tool to connect to a distant LAN ROSE switch and request a HEARD list of all the users of that area who have been active in the last few hours. This gives us a vehicle to discover who is operating packet beyond the horizon. Once you have connected to HEARD at the distant switch, press <enter> a second time and the HEARD list is initialized and the area USER list will be sent to your station. The display you receive will be of the most recent HEARD stations to the oldest HEARD. The list generally covers the last three hours. The time may vary in accordance with the amount of switch use.

Refrain from downloading HEARD lists during "high traffic" periods. There may be heavy traffic at the other end of the HEARD path. The HEARD feature can become a very useful feature if it is not abused.

When you receive **END**, issue a normal disconnect.

For Glynn, WB4RHO, near Dothan, Alabama to connect to HEARD near Barnwell, South Carolina the following connect would be used:

C HEARD V WB4RHO-1,404592 <enter>

The Fundamentals of The ROSE

ROSE uses the *Country* Code, the *Area* Code, and the *Local* (telephone prefix) Exchange numbering system. This topology is often referred to as **CCITT** numeric addressing. CCITT is the Consultative Committee International Telegraph and Telephone, an international organization that sets worldwide communications standards.

The ROSE code as received is defaulted to the Country Code for the United States. This code is set to 3100 for the USA and is not seen by the users, nor is it necessary for the user to input this Country Code unless there is a crossing of boundaries to another country. The only concern that we users should have is the Area Code (3 digits) and the Local Exchange (3 digits) numbers (total 6 digits).

In the United States the Area Code and Local Telephone Prefix numbers are used to address the ROSE system. For instance, the Area Code for central and south Georgia is 912. One of the 3-digit exchange numbers for Macon, Georgia is 781. Thus, the address for the local 1200 baud ROSE switch is 912781.

There is a diversity of ROSE switches, network nodes, and digipeaters in use across our country. Where the ROSE switch exists, there are many combinations of routing schemes that can be used to build Virtual Circuits (VC).

As the popularity of the ROSE grows, so will its use. For now we are making use of the nodes and digis that are still active along the routes.

To make use of these other packet transfer mediums we can include the nodes or digipeater calls in the routing tables of the ROSE switches. As the nodes and digis are phased out or replaced by ROSE switches, we can easily configure the ROSE switches so they can recognize the new switches and make use of them in the throughput paths. Simply put, this means we can program around the network nodes or through them, as we wish. If the SYSOP is truly familiar with the configuration of the ROSE, he/she can write the configuration so well that the user may never be aware that nodes or digis are even a part of the path.

The Network "Node"

P R U N I C CQ is *not* an acronym. However, each letter stands for each of the user feature commands within some of the network nodes. You can arrange them in any combination that makes it easy for you to remember them. They are as follows:

P—Returns a list of the node **P**arameters. The **PARM** command is sent while connected to the node.

R—Returns a list of the current node **R**outes that are known to the node.

U—Returns a list of **U**sers on the node at the present time. This command feature will also show any "nested" user CQs.

N—Returns a list of all **N**odes that are known by the connected node.

I—Returns the Information or the type of node that is being used.

C—Accompanied by the callsign of another node, or station, enables connects to another station or to another node using the level-three packet protocol.

CQ—Allows a connected user to issue a CQ from a distant node, which will enable stations in the distant LAN to connect to you at the distant node, as though the distant node was your own "remote" TNC. This command provides the user with a means to attract other stations in the area to a distant node. Should a packet station in the area of the node see the "X4XXX-15," it can connect to the displayed call, and a link will be made.

The feature commands that you have just read are the "most used commands." Let's cover these commands in more detail and study the usage in a real-time environment.

The most used (and often abused) of these commands is the NODES command. This command feature can give the user a list of the other nodes that can sometimes be reached by the connected node. The "abused" part of this command is that sometimes a user will attempt to pull a node list from a distant node using a poor path. The end result is that the users of the LAN at the distant node location are bombarded by the nodes lists that are retrying over and over. These nodes lists sometimes contain more than 50 calls and aliases. The task of remembering all the routes to the other LANs in your area is something of the past. Within the Network Node Controller (NNC) EPROM, a connect to a distant station can be reduced to as few as three callsigns to remember.

The network EPROM has embedded into its code an algorithm which allows it to update the routing tables about once an hour. The updating is achieved through communications with other NNCs. The nodes "swap ideas and routes," so to speak. One network EPROM node can keep up with as many as 80 other nodes through the network of NNCs which surrounds each of them.

CONNECT Command: With the CONNECT command we begin the scenario. The nodes mentioned here no longer exist, but are listed herein as an aid to explain how they are used to establish a connect using the Net/Rom and TheNet formats.

My brother Mike lives near Anniston, Alabama, about 300 miles southeast of my former QTH. Let's say Mike, N4NAU, monitors ALA5 on 145.05 MHz in the Anniston area. I want to connect to Mike, so

I connect to a local Nashville node, NASH2, on 145.01 MHz, and tell it to connect to MAURY node near Columbia, Tennessee. MAURY appears in the NASH2 node list. Thus, NASH2 knows how to connect to MAURY.

There are other nodes between NASH2 and MAURY. This is why the nodes are often talking to each other. They are updating the node lists of one another with the most recent routing information, and sometimes noting erroneous routing data that was gathered during a band opening.

Next I tell the MAURY node to connect to node HSV1 at Huntsville, Alabama. Again, there may be many nodes between the MAURY node and the Huntsville node.

The HSV1 node list display tells me that it knows the way to JVL01 near Jacksonville, Alabama. True, I don't know the route to JVL01. All I know is that when I pulled the node list, I saw the node for Jacksonville listed in the display. There are several nodes between HSV1 and JVL01, and in addition, there is a gateway from 145.01 to 145.05. The 145.05 port of the JVL gateway is defined in the node list as JVL05. I issue a connect request to JVL05, and in a few minutes I'm connected to the JVL05 node on 145.05 MHz.

Finally, I issue a connect request to Mike's callsign:

C N4NAU < enter >

In less than a minute I am greeted with this message:

*** CONNECTED to N4NAU

My next input is to send a greeting to Mike, or a simple Hello Mike, etc. If Mike is not home, or for some other reason he has turned off his packet station, I will receive the following:

FAILURE WITH N4NAU

USERS Command: I remain connected to node JVL05 after I receive the **FAILURE WITH N4NAU** message. This allows me another chance to use the USERS feature/command. If I type a **U** and < enter >, I could possibly find another station in the Jacksonville area listed in the USERS list in the node. Even better, I might even discover a nested CQ in the node JVL05. Either type user connect can be identified by an UPLINK or a CQ by the callsign in the received listing. If there are other stations connected to the node in the CQ mode, you will receive this reply:

You can connect to CQ callsign by issuing C X4XXX-15

where X4XXX is the callsign of the CQ mode connected station.

The USERS command allows us to see the callsigns of other users of the node and the callsigns of the stations to which they are connected. You may see something like this when you issue the USERS command to the connected net type node:

USERS JVL05:X4XXX-5 [net type node]

Uplink (N4QLG) Downlink (N4QLG: > BHAM) CIRCUIT (HSV1:JVL05: > K4ABT-15)

CQ Command: If there are no users listed or CQs nested, then I can issue my own CQ. This is called a "wild card" call, and it will display my call at the JAX7 node with a CQ by my callsign. Quite likely, it will have my call displayed as "-15." The SSID of "-15" is usually how the callsign is displayed at the far end, if you began the connect path through the nodes with a callsign and no SSID. *Note:* The CQ command is available in most versions of the net type node EPROM. *This is a sleeper command of which very few node users are aware.* When implemented properly, it can prove to be a most useful tool in making contacts with stations in other LANs that were unheard of before. Here is how it works.

I send a CQ and a short message in this manner:

CQ < space > **CQ** Buck in Gallatin, Tennessee with traffic

Up to 77 letters, spaces, and characters total length can be used in the message. Note there is a space after the first CQ. *The space is absolutely necessary for the CQ command to work properly.* After displaying my call to the distant LAN, the CQ stays active for 10 or 15 minutes, depending on the setting of parameter number 15 on the node. Anyone who sees the message can connect to the callsign they see displayed at the distant node end. The callsign will have a translated SSID attached (usually SSID -15). The translated SSID is added by the net type node and should be used by the station attempting a connect.

Circuit (NODE:callsign-?)   ..CQ (KK4XXX-15)

You can connect to CQ callsign by issuing **C KK4XXX-15**. (KK4XXX is the callsign of the CQ node connected station.) This will give you a vehicle to discover the callsigns of stations in the distant LAN.

INFO Command: The INFO command is invoked by issuing an **I** to the connected node. This command will return the SYSOP's call and SSID, plus the alias callsign of the node to which you are connected. Many nodes will display the version number of the node firmware. In addition, you may receive a long message giving the node location, power, frequency, etc.

If you were to connect to ABT1 and issue the **I** command, you would perhaps receive the following message:

K4ABT-1:ABT1 >

Located near GALLATIN, TENNESSEE
SYSOP is BUCK
SEDAN NODE
145.01 MHz port of gateway to ABT5 on 145.05 MHz

NODE Command: While connected to the network EPROM node, issue the NODE command and you will receive a list of all the nodes that can be called from the connected node. This will be a list of the nodes in its routing tables. The list will appear similar to the following:

MCN5:K4ICT-5 Nodes

ABT05:K4ABT-2	ABT1:K4ABT-1	ABT5:K4ABT-5
ABT6:K4ABT-6	ABT-7:K4ABT-7	ABT8:K4ABT-8
ABT9:K4ABT-9	ABT10:K4ABT-10	ABT11:K4ABT-11
ABT12:K4ABT-12	ABT28:K4ABT-3	ABY1:W4MM-1
MGY:W4AP-2	MGY4:W4AP-4	PERRY5:K4ICT-11
QSO:WB4EDZ-1	RHO1:WB4RHO-1	RHO2:WB4RHO-2
TCL1:W4KDP-1	TCL3:W4KDP-3	WGA5:KS4C-5

After you've used the NODE command, save the list that you get and combine it with other node lists, and soon you will have a routing table list of your own. (Definitions of the NODE commands are listed in table 3-1.)

ROUTES Command: Here is an example of the routing table that was downloaded from GARDS node ABT8. Connect to ABT8; then enter an **R**.

Suggested Parameters for TheNET Version 2.10

Function	User Node	Backbone Node
1 Minimum quality for auto update	50	50
2 HDLC channel quality	50	203
3 RS-232 channel quality	203	203
4 Obsolescence count initialize value	3	3
5 Obsolescence count minimum for broadcast	4	1
6 Nodes broadcast interval (in seconds)	1800	1800
7 FRACK (in seconds)	8	1
8 MAXframe	1	1
9 Link RETRIES	10	10
10 Validata callsign 0 = NO & 1 = YES	0	1
11 HOST mode connects	0	0
12 TXDelay (X 10 milliseconds)	35	35
13 Broadcast via port: 0 = disable all; 1 = enable radio (0); 2 = enable RS-232(1); 3 = enable all (0&1)	3	3
14 Pound node propagate	0	0
15 Connect command enable 0 = No & 1 = Yes	1	0
16 Destination list length	100	100
17 Time to live initializer	7	1
18 Transport timeout (in seconds)	200	200
19 Transport RETRIES	2	2
20 Transport ACK delay (in seconds)	1	1
21 Transport busy delay (in seconds)	180	180
22 Transport window size	2	2
23 Congestion control threshold	4	4

EPROM Parameters	Users	Backbone
24 No activity timeout (in seconds)	7200	7200
25 P—Persistance	64	255
26 Slot time (X 10 milliseconds)	20	1
27 Link RESPonse time (X 10 milliseconds)	50	20
28 Link T3 timeout [CHECK] (X 10 milliseconds)	65000	65000
29 Station ID 0 = No & 1 = Yes	1	0
30 CQ broadcasts 0 = No & 1 = Yes	1	0
31 HEARD list length	20	20
32 Full Duplex 0 = No & 1 = Yes	0	0

Table 3-1. Definitions of the node commands and what are considered by some SYSOPs as the best default parameters.

ABT8:K4ABT-8 > Routes:
> 0 K4ABT-11 192 44
  0 WB4EDZ-1 0 1
  0 K4ABT-9 100 2
  0 K4ABT-12 100 11
  0 K4ABT-5 100 20
  0 WB4RHO-1 192 34

PARAMETER Command: The final command for our discussion is the PARAMETER command. It is downloaded from a node by entering a **P**.

NOTES

Chapter 4

Conference Nodes

The **CONVERSE** node (sometimes called the CONFERENCE node) attracts many users, and it is a natural for the Local Area Net (LAN) frequency. Not only can it be used as a local round-table packet session, but it doubles as a digipeater. The outgrowth of this node can be used for networking in a VIA type connect, or it can be utilized as a round-table or for holding a net. The CONVERSE supports multiple on-line conferences or round-tables between keyboard users. It can be connected to in the same manner as any network node.

The code is burned into a 27C256 EPROM which fits into the TAPR TNC-2 or clones such as the MFJ-1270B, MFJ-1274, PacComm Tiny 2, and DRSI DPK-2.

The callsign of the CONVERSE node will appear in the net type node tables just as any other node will appear. It allows connects in the same manner as The-Net type nodes. The difference is that the CONVERSE will only digipeat; it will *not* allow calls to be made "from" it (only "to" it). The Mini-CONVERSE is a "Terminal" node. In other words, it will not operate in level 3 packet mode.

The CONVERSE node operates as a stand-alone system or it may be used with a cluster of nodes.

Initializing

The node sends out update broadcasts to inform other nodes that it is active. The CONVERSE node can support 256 channels of communication; each channel can have two or more users. The operating parameters are set in the firmware and are not available for easy changing by the SYSOP. These parameters are located in the same place as the parameters of TheNet firmware and can be modified at the time the EPROM is configured.

Each user in a round-table receives all the information from every other user in the net or round-table, unless another station uses the /Message "callsign" to send a private message. Each packet may contain the callsign of the originator, and it will display at each receiving screen. Another user can switch channels and all users on the channel will be informed that "station X4XXX has changed to channel [n]." If the channel to which station X4XXX moved is not occupied, an **INVITE** can be issued to other users. The INVITE can be to users on other channels as long as the correct callsign is used in the INVITE. If the channel to which station X4XXX moved is occupied, other users already on the channel are notified that "X4XXX has joined us." The /**I** [callsign] feature will be sent only to the station that is identified in the [callsign].

Zero Is The Default or Calling Channel

When first connecting with the CONVERSE node, the user is defaulted to Channel zero (0). Channel 0 is considered to be the calling channel. The callsign of our local CONVERSE node is WB4EDZ-6 with an alias of QSO (QSO on 145.650 MHz).

Connect to **QSO** and immediately issue a /**H** or /**?** for the HELP menu. If you are already familiar with the CONVERSE MENU, then issue a /**W**. The /W will set the node into action to recognize your connect. If you connect to the CONVERSE node and make no other entry, then you may not receive any response. A timer in the node sees that no entry has been made, and at the end of approximately ten minutes you may be disconnected.

Here is how we use the CONVERSE node in conjunction with the Sumner County Red Cross operations.

After connecting to the CONVERSE node, type /H or /? and <enter>. You should receive the following menu:

*** CONNECTED TO K4ABT-6 (SCRC)

Commands can be abbreviated:
/HELP—Help
/EXIT—Terminate CONVERSE session
/BYE—Terminate CONVERSE session
/QUIT—Terminate CONVERSE session
/CHANNEL n—Switch to Channel n
/INVITE user—Invite (user call) to join your channel
/MSG user text—Send text to one user only on the channel
/WHO—List all users and their channel numbers

WELCOME TO THE SUMNER COUNTY RED CROSS COMMAND POST AT GALLATIN, TN
***

The *** is the prompt for further input, so you might enter

/WHO

Note: The first line of the menu indicates that commands may be abbreviated. Therefore, you may enter /W and obtain the same information as you will with /WHO. The reply to your /W will appear similar to the following listing:

User	Circuit	Channel
KC4NEH	K4ABT-9,615822	0
N4SSB	K4ABT-9,615822	24
KD4BNW	K4ABT-9,615822	24
KO4GF	K4ABT-1,615528	0
N4UBR	K4ABT-1,615528	24
K4ABT		0

***

KC4NEH issues the following command:

/C 24

At N4SSB, KD4BNW, and N4UBR the following appears on their screens:

*** KC4NEH signed on:

This message appears on the screens of all stations left on channel 0:

*** KC4NEH changed to channel 24:

Here is a typical QSO as seen by the channel 24 users:

<N4SSB>: Hey, Richard, I see Perk joined our QSO.

Richard sends a private message to Bill, N4SSB, by typing:

/MSG N4SSB Hey, Bill, don't forget to remind the guys of the meeting this Thursday night at the Golden Coral.

Only N4SSB receives the message. It appears as follows:

<*N4UBR*> Hey, Bill, don't forget to remind the guys of the meeting this Thursday night at the Golden Coral.

The asterisks (*) inside the less-than/greater-than symbols tell Bill who the private message is from. In this case the message is sent from N4UBR to N4SSB. All messages you send starting with a / (slash) symbol are directed to the command interpreter of the CONVERSE node. Furthermore, you may invite other users from other channels to join your channel with the "/Invite command" /I "CALL" <enter>.

With this short explanation I feel we have covered how the CONVERSE node overcomes the problem of not being able to have roundtable discussions in packet radio. This is only the beginning of the fun

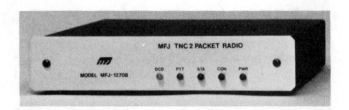

The MFJ-1270B can be converted to ROSE switch or node use. (Photo courtesy MFJ Enterprises, Inc.)

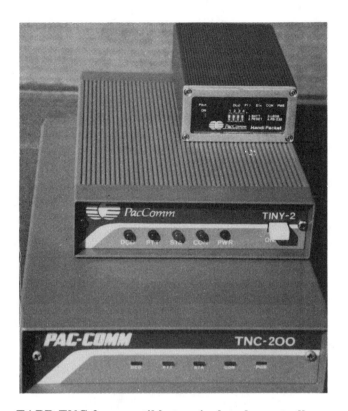

**TAPR TNC-2 compatible terminal node controllers.
(Photo courtesy PacComm Packet Radio Systems, Inc.)**

things that can be enjoyed with this "user friendly" CONFERENCE/NET node.

Just one final caution: It is acknowledged that the QSO/CONVERSE nodes do not have a place on the through-put frequencies, since they can truly bring a LAN to its knees while in use by several packeteers.

If you wish to use the CONVERSE node as an adjunct to your emergency, weather watch, or packet spotting network, then make every effort to place it on a frequency that will not interfere with other modes of packet operation. Place the CONVERSE node on a frequency that *will not interfere* with other LANs, and you will have many hours of round-table fun and excitement with the CONVERSE node, not to mention its value in the event of adverse circumstances and emergency situations.

NOTES

Chapter 5

Bridges and Gateways

Throughout the southeast we have a rather large keyboard-to-keyboard network in place. This network is accessible from many different frequencies and baudrates. The ROSE network is used primarily as a long-range communications system whereby one station in, we'll say, south Alabama can converse with friends in Illinois.

There are times, however, when the network is requested for emergency traffic. When this happens, or the signal is given from an Emergency Operations Center (EOC) along the network, this network comes under the control of the EOC that has requested the use for the specified area.

Because this network services more than 20 states, it is recognized that the system cannot use only one frequency to cover such a wide area. We therefore have employed the available frequencies in certain areas to circumvent colliding with LANs, BBSes, or packet spotting systems. The network was built with the idea that we can and must be good neighbors, so there has been a lot of extra expense in making it "fit" the available spectrum.

The important factor to remember is the target station to which we are connecting. Although the network east of Nashville, Tennessee operates on 145.05 MHz, the network west of Nashville is on 145.07 MHz. With this in mind, we have a minor problem with the traffic coming from one direction and traversing onto the frequency in the opposite direction.

Figure 5-1 will enable you to understand how easy

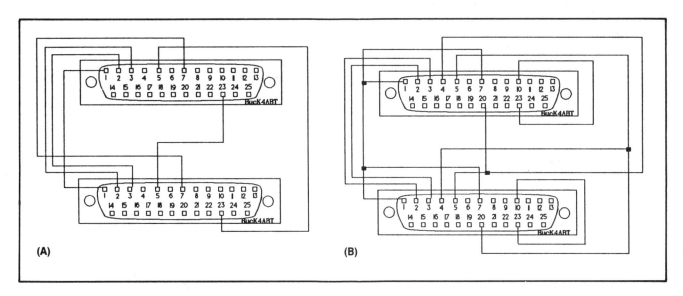

(A)

(B)

Figure 5-1. (A) This cable interfaces two TNCs that are configured as ROSE switches. (B) This drawing illustrates how two nodes may be interfaced to annex LAN nodes to a neighbor node on the backbone frequency.

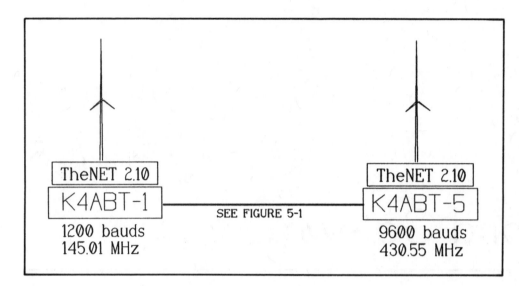

Figure 5-2. This drawing shows how a LAN node allows access to the 9600 baud backbone frequency. As illustrated here, the interface cable in figure 5-1 "bridges" the two nodes to join the two frequencies and baudrates.

it is for the system to complete the connection between neighbor frequencies. Furthermore, we've included additional information in figure 5-2 to enable the use of a "bridge" to and from a high-speed backbone. In other words, the user need only have one TNC with only the local LAN baudrate in it to connect to another station many miles away through many different frequencies and bauds.

These gateways are only part of the networking complex that allows the LAN users to enter and exit to other frequencies with little or no effort. In some instances the user never realizes there have been any diversions along the path(s).

Near my former QTH there are two such gateways, or bridges. The first gateway is the bridge between 145.05 MHz and 145.07 MHz. These are N4SSB-5 /615452 and N4SSB-7,615847 ROSE switch gateway combination.

N4SSB-5 node is ported from 145.05 MHz to 145.07 MHz via an RS-232 cable. A user in Georgia can establish a connect on the 05 LAN and direct the connect to a station in Missouri. The connect automatically connects to the Missouri user through some 20 or more switches, yet the originating station in Georgia has no knowledge of what route was used or what baudrates were employed in making the connect path. The reason is because the ROSE ops along the route installed the routing to the neighboring switches and included the area codes and office codes for the surrounding call prefixes.

We will cover more on the addressing schemes used to accomplish this kind of routing in the chapter on the ROSE network.

Chapter 6

Packet Video

We've come a long way in digital communications, and furthermore, we have developed faster speeds and better means to transfer, display, and store binary data—not just any data, but high-resolution, color-picture data. *Pictures* via packet? Yes!

Creating and Sending Video Via Packet

Little did HAL Communications realize in 1982 and 1983 what they were introducing to the amateur radio community when they released the Electronic Mailbox. The TRS-80's (I, II, and IIIs) and the Apple II and II + computers were making their way into the ranks of amateur radio. These computers were the powerhouses of RTTY because they were faster and

easier to interface to the new mode of traffic handling via the mail storage option (MSO). Two of the major suppliers of the MSO systems were HAL Communications Corporation and Mactronics Inc. These MSO mailboxes and BBSes supported speeds from 45 to 110 baud ASCII.

In those days we could upload and download any BASIC program or ASCII file as long as it was saved in an ASCII format. That was only a small part of what you could find on the MSOs. There were the ever-familiar RTTY pictures. These pictures were letters and numbers arranged in such a way that when printed out to the long printer paper (in most cases, rolls) they would present the recipient with a very good picture or scene. There were not many ways to display these masterpieces, since the sheets of paper were usually several feet long, or it required placing the sheets

The HAL PCI-4000/PC-CLOVER is a CLOVER modem for HF data communication. (Photo courtesy HAL Communications Corp.)

This is a photo, not retouched, of the screen just after the image was received via packet video by the author.

side by side to form a wider tapestry. In addition, in the days before disk storage media we only had tape, and in some cases this was not magnetic tape. It was a perforated paper tape.

For a moment imagine, if you will, how much room it would take to store one picture which sometimes had as many as 20,000 letters and numbers to compose the illustration or picture. Once we received the complete picture (after several tries) and saved it to disk, we went about the long task of printing the picture. The advent of the MSOs and disk storage media

was a boon to the picture collectors, because they could sometimes store ten pictures on a single disk. This in itself was a bit expensive in those days, since the first disks I purchased were $7.00 each. I could buy a box of ten disks for $50, but I still thought twice about which picture I would download from the MSO.

We've Come A Long Way

We've come a long way in digital communications since then, and furthermore, we've developed faster speeds

A favorite among packet picture passing enthusiasts is this Pitts Special.

The American eagle is a packet favorite for passing in 256 colors.

and better means to transfer, display, and store pictures. In addition, double-sided, double-density disks with labels and shucks can be purchased for as little as 25 cents each!

Picture Transmitting

A new method of picture transfer, display, and storage has emerged which makes the old way look like hieroglyphics by comparison. This new system has improved to a point that makes it possible for us to transmit high-resolution, color pictures via packet. Depending on the type of monitor and driver card, we are able to transmit and view CGA, EGA, and VGA pictures. Even the pictures that are more detailed and contain 800 × 600 pixels with upwards of 256 colors are easily passed via packet. The picture will display as it is received on the screen, in color, and it will automatically save to the disk with the same title as it had on the originating station's disk. And, the picture will only need to be transmitted one time, since the AX.25 packet radio technique treats the transfer of a digital image as if it were any other binary data file. The error-free capability of the AX.25 protocol is the key to this very effective means of digital image transfer (DIT).

We now have access to dot-matrix printers that allow us to print the picture in near photographic quality. We also have programs which will let us add other graphics or print messages within the picture.

This brings us to the subject of the many letters I receive from readers who are into collecting packet pictures. The building of the pictures for RTTY was a skill unto itself, while the development of the pictures was somewhat easier because the pictures were composed using the keyboard of the terminal.

Although easier to transmit, receive, display, and

This image was received and displayed on the screen as it was being received, and the file was stored to disk at the same time.

save, the generation of high-resolution pictures for packet becomes a bit more complex.

Picture Creation

We can depend on others to develop these Rembrants, or we can purchase the interface to build our own pictures. This places the prospective graphics and picture developer in a situation in which he or she can use one of the many drawing or paint programs to build the pictures. These pictures can sometimes be an expression of the artist's wit and wisdom, or just a "stick" drawing. With the new scanners which pick up a high-contrast picture and save it in a binary format on disk, we are able to make copies of some already hard-copy elements.

The latest Logitech Scanman Plus scanner provides some amazing results when compared to similar video scanners of only a few years ago. An additional feature of this new Logitech family of scanners is the support for real optical character recognition (OCR). These scanners are in the $150 to $200 category and represent a quantum leap in the user-friendly capture of pictures that can be sent via packet radio.

Picture Digitizers

Another way to capture or generate pictures is supported by the digital converter, or "digitizer," as they too are beginning to reach an economical level for the end user. The digital picture converter can be used with your home video cassette recorder (VCR), or it can have composite video input. This enables us to feed video from a small black-and-white video surveillance camera or from a color video camera/recorder combination.

The quality of the picture is a product of lighting, camera quality, and video-converter gray-scale capture. The software, which is part of the video to binary converter, is important because this allows the user to add dither or to increase or decrease the level of contrast.

There are several makes and models of hand-scanners and digitizers available to the packet picture collector. I have used both the hand-scanners and video digitizers. Either will enable the capture of any picture, large or small, and save it to a CGA, EGA, or VGA file. I use all three: the Logitech Hand-Held, the VGA ComputerEyes, and a Picture Perfect digitizer. The software for the Picture Perfect digitizer is by Bob Slomka, WD4MNT. Bob is the author and developer of many packet radio terminal programs for the PC and compatibles.

With this software we are able to save the picture into one of several different formats, and one save even allows raw data to be saved so the user may later load the raw data and save it to another format so that it may be used with SSTV, packet, or saved into a .PCX format. The latter save allows the user to load it into the drawing package and view, color, add text, flip, turn, reverse, or manipulate the picture in so many ways that it may not even resemble the same picture when the changes are complete.

Old Glory is another very popular image for passing via packet. Note there are no streaks, as packet pictures are received and displayed error-free.

After I have generated the picture with the Picture Perfect digitizer, I will sometimes add a few "refinements" of my own, or I will include my callsign and address in an area of the picture where it will be legible. To do this I must have a drawing or paint program which allows me to pick the picture off the disk and load it into the buffer of the paint program, where I add the enhancements. When I am finished with the color adding, color enhancing, or text adding, I can save the picture to the disk again.

A standard CGA formatted picture, after the software compression, can be as small as 2K up to about 17Kb. The largest of the VGA pictures will be about 50K. We find that storage of the CGA pictures will allow about 25 to 30 pictures to be stored on a single floppy disk.

With the Picture Perfect video digitizer and related software, these pictures can be generated right inside your PC or clone. The reason I emphasize this point is because the Picture Perfect software enables the user to save a screen of video into any one of several formats. The program also permits disk storage of the picture data to a "raw" video format for use at a later time. Raw video permits the user to reload the data to Picture Perfect later and save it out into any of the other different formats—e.g., SSTV, FAX, etc.

The author has added another touch of genius by combining the power of his MFJ-1292 Picture Perfect video digitizer software and allowing the user to capture and save a picture in the 8-level facsimile format. The formats of this Picture Perfect video digitizing system do not stop at the FAX mode(s). Included in this package are the user capability to save the pictures into several slow-scan television modes, PCX format for use with the PC PAINTBrush drawing software (trademark, copyright ZSOFT), and into a special Packet Picture Format of either CGA, EGA, or VGA, and that's just for openers. The hardware and software of the MFJ-1292 Picture Perfect make it a very useful and rewarding add-on to this system.

Other picture digitizers, such as the COMPUTEREYES digitizer, enable the capture of color VGA pictures in many formats that can be converted into packet picture formats. Consult your computer accessories vendor for more information about these color VGA digitizers.

NOTES

Chapter 7

How To Build Switches, Nodes, Gateways, Dual Ports, Backbones, and Trunks

Not only do I get requests for the "how to do it," information but I also get requests for the actual drawings and illustrations that describe the ways we can interconnect switches, nodes, and gateways. Recently, we have begun to receive a lot of requests for 9600 baud information that describes how the backbones and trunks are configured. The demand for this kind of information has increased dramatically. It takes a long time to turn out individual drawings, and it is not an easy task. My CADD5 is not the latest version, so it may respond a bit slower than new versions.

Are They CONFERENCE or CONVERSE Nodes?

Both! In addition to the node and gateway interest, there is an upswing in interest regarding the backbone and CONFERENCE nodes. The concern for the CONFERENCE nodes is generally centered around how to include one on the LAN frequency. I'm not trying to set policy for the LANs coast to coast, but it would be good to consider how the CONFERENCE node will be used before implementation on an active LAN frequency. I can assure the prospective CONFERENCE node SYSOP (as I have before) that a CONFERENCE node, when active, can bring a LAN to its knees.

In this chapter we will cover the manner in which a CONFERENCE node can be used in conjunction with a gateway, although I do not recommend this application of the CONFERENCE node. Only if the frequency is one which has a limited (small) number of users can a CONFERENCE node and a gateway node survive. A CONFERENCE node should be placed on a seldom-used frequency and away from gateway nodes. As many of the SYSOPs who are reading this know, a CONFERENCE node will find a way to the node lists of neighboring nodes. Once this happens "the fat's in the fire."

The future of digital communications lies in digital signal processing. The way of the future is happening now in the form of the DSP-2232 multi-mode data controller. (Photo courtesy Advanced Electronic Applications, Inc. —AEA.)

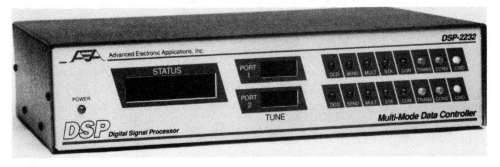

Remember this: Every packet that is sent to the CONFERENCE node will be sent to every station connected to the round-table, and every packet station will receive an "ACK" packet. Likewise, the CONFERENCE node will send an ACK to every station connected. Now that makes for a busy frequency! In other words, if four stations are connected to the CONFERENCE node, then eight packets are generated by one packet sent to the node.

If the connection(s) are through a gateway or a second node, then consider multiplying the just-mentioned packets by four. This means 32 packets have now been generated by one packet sent. What if just one packet is missed in all this jumble and a second try is needed? Bingo! You've got it—mass collisions.

There Are Other Nodes

When a system begins to grow, the user loses sight of how and where the packets go after they leave the Local Area Network (LAN) frequency. The SYSOP is the only one who really has some notion of what happens to the packet(s) outside the LAN. Here is where the plot seems to thicken. When packet was in the embryonic development stages, it was no problem to keep track of how the data was handled. In the past two years, however, there has been such a large increase in packet use that the switch and node SYSOPs are having to improvise just to keep their LANs going, or as in a few cases, they have ceased to be SYSOPs altogether. The problem lies with the lack of available information that would enable the SYSOP to upgrade his system.

I can hear a few of the SYSOPs ask themselves, "Is Buck kidding?" No, I'm not. My incoming mail has spoken loud and clear! Information that we SYSOPs take for granted is not in great supply in most regions of our country. Questions about ROSE switches, nodes, gateways, node complex, backbones, and trunking are but a few of the topics we will cover in this chapter. I'm not omitting the digipeater, but why cover something that all TNC manuals have covered many times over? Let's outline what we are about to cover so that you can determine which one of these items will best fit your needs or applications.

Building and Configuring The ROSE AX.25 Switch

To begin building the first of many ROSE switches we initially obtained the latest version of the ROSE code and support files. The files are somewhat larger than a 360K disk can hold, so the author (Tom Moulton, W2VY) used the PKZIP © method to compress the files into a smaller size so they could be stored on an inexpensive, lightweight, and easy-to-mail 5¼ inch 360Kb disk. **Do not try to unzip the files onto the disk that contains the original zipped files!**

I moved the zipped files to my hard disk and ran the PKUNZIP.EXE program on the two main files to convert the zipped contents to readable and usable files. The documentation and README.DOC text files were the first files that I put to use. I printed the RZSYSOP.TXT and RZUSERS.TXT files so I could study them. I never finished reading the USERS.DOC, but I did read through the SYSOP documentation.

MAKEPROM.EXE:

The next action was to run the MAKEPROM.EXE (on an IBM PC compatible) program and use the TNC2.OVR to set the parameters to match those needed to configure the EPROM for use in a TAPR TNC-2 clone. My first attempt to make an EPROM was altered in the ADDress, because I did not use the United States telephone ID code (3100) before the area code and local exchange.

I erased the EPROM (27C256/12 volt) and began once more. If I had paid closer attention I would have noticed that the MAKEPROM program already had the example for me to follow installed into the ADDress location in the display. The final ADDress was "3100912929" with 912 being the area code and 929 the local exchange. The CALLsign and ADDress were K4ABT-5/912929. Our ROSE network was to consist of two switches, so I burned a second EPROM with the CALLsign and ADDress K4ICT-5/912781.

Installing The EPROMs in The TNC

The TNC-2 and clones use the socket at U23 for the systems EPROM. I've learned to first check the TNC to ensure that it is in good working condition. Occasionally, you may encounter an older version of the TNC-2 which will not function as a ROSE switch. To use a TNC-2 as one port of a gateway, you must follow the instructions in the SYSOP ROSE manual to wire the RS232 port connectors. Install the correct jumper inside the TNC (see figure 7-1).

Install the EPROMs and be sure no pins are bent under the IC or outside the socket. Before replacing the TNC cover, test the TNC by powering up to be

Figure 7-1. This minor modification enables flow control when two or more TNCs are connected back to back, or by a diode matrix cluster. Add the jumper as shown. The TNC-2 and clones manufactured after 1990 have this jumper already installed.

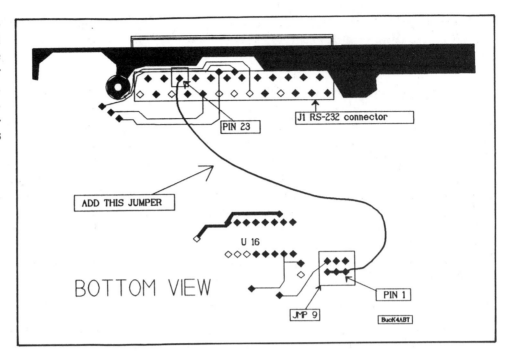

sure the CON and STA LEDs are cycling (flashing back and forth) at a one second interval.

First-Time Confirmation Of The ROSE Switch

Initializing the ROSE switch is the first action of the SYSOP. Before going to the site to install the ROSE switch, we *must* be sure it is functioning as it should. With all cable connections and interfacing completed, the power is applied to the TNC/ROSE switch. The CON and STA LEDs will both light up and stay lit for about 5 seconds. They should begin blinking, alternating between the CON and STA LEDs at 1 second intervals. This is the indication the SYSOP is looking for in a properly operating ROSE TNC.

If for some reason there is an error in the TNC, one of the following will occur:

A. LED flashing will not occur;
B. All LEDs will light up and stay lit with no flashing; or
C. There will be random flashing of all lights.

Any combination of the above occurrences, other than the one second interval flashing of the CON and STA LEDs, indicates a problem in the ROSE TNC, and further trouble-shooting is necessary.

Preparing The CNF and TBL Files

By following the examples in the ROSE SYSOP's manual supplied with the files in the original disk, we can build the configuration files for our switch(es). Remember to identify each switch that is a neighbor switch on the same frequency, or any switches that are addressed through the RS232 ports (PORT 1). Use the extension **.CNF** to identify the TEXT version of each switch configuration file(s). The last command in the (ASCII/text) .CNF file is a **WRITE TO:** command. Here I use the same file name as the .CNF file in which I'm currently working—with one exception. That exception is *use the extension* **.TBL**.

Here is how a .CNF file might appear:

Default L3W 2
Default TimeOut 500
Default MaxVC 30
Default Port 0

PASSWORD 615666.PWD

This DNIC 3100 United States of America

This Node LafayetteTN7
Address 615666
Call KC4NEH-3
Digi KC4NEH-4

Coverage
615633
END

Userport 0
Text
* < ROSE.INS
$
WELCOME to the ROSE @ LAFAYETTE,
   TENNESSEE
$EOF
END

Node GallaTN7
Address 615452
Path N4SSB-3
Port 0
END

Node TTUTN7
Address 615372
Path WA4UCE-3
Port 0
END

Node CookevlTN7
Address 615526
Path KO4GF-3
Port 0
END

Node CumberlandcityTN7
Address 615289
Path WA4TLZ-3
Port 0
END

Route to Nodes GallaTN7
Calls for
615452 615822 615451 615555 615824 615264 615325
   615386 615860 615865 615672 615741
END

Route to Nodes TTUTN7 CookevlTN7
Calls for
615372 615526
615
* < AL.NPA
* < GA.NPA
* < FL.NPA
* < SC.NPA

* < NC.NPA
* < W8.NPA
* < W3.NPA
END

Route to Nodes CumberlandcityTN7
Calls for
901642 615289
901 601 502
* < W9.NPA
* < W0.NPA
* < W5.NPA
END

WRITE 615666.TBL
QUIT

The above .CNF text file is generated in the F10 editor of BUXTERM.EXE by the SYSOP. It is then saved as 615666 with an extension of .CNF.

Once the file is complete, save it to disk with the ALT S command, using the title **615666.CNF**.

The .TBL hexadecimal file is created by the CONFIGUR.EXE (C.EXE) from the .CNF file by the **BUXTERM F9** key. Afterwards, a new title will appear in the directory with the file name **615666.TBL**.

This has proven to be a good way to keep the files alphabetized (numerically) in the BUXTERM directory. After the file has been converted to a .TBL file with the F9 (C.EXE see BUXTERM.DOC) key in BUXTERM, it should have the same file name as the original .CNF file except the extension should be .TBL.

Configure The New ROSE Before Moving It To The Site

When a ROSE switch is first activated, it has only one resident access or entry module. Burned into the EPROM of the virgin ROSE switch is the **LOADER** module. The idea is to operate the new ROSE as if it were in the environment or site where it is to be used. Let's imagine the switch call is K4ABT-1 and the ADDress is 615386. The SYSOP makes his/her first connect to the new ROSE switch in this manner:

C LOADER V K4ABT-1,615386 < enter >

The switch response is similar to the following:

cmd:*** CONNECTED to LOADER VIA K4ABT-1,615386
Call being Setup
Call Complete to LOADER-0 @
3100912987
ROSE X.25 ROSE switch Ver. 9#####
by Thomas A. Moulton W2VY
OK

Since this is the first time the switch has been used, we *must first send* **BOOTER.LOD** to the LOADER module. Employing the BUXTERM terminal program, send BOOTER.LOD to the new ROSE switch by holding down the ALT key and pressing the R. This invokes the directory that should contain the .LOD, .CNF, and .TBL files. Use the arrow keys and move the highlight to the BOOTER.LOD file and press <enter>.

The short file will be sent to the switch. In a few moments the switch will DISCONNECT. The new switch has now set the register pointers into the correct hierarchy, and the ROSE is now ready to be configured.

Once again we connect to the LOADER in the same fashion that we used for the first connect:

C LOADER V K4ABT-1,615386 <enter>

Again we receive the same connect sequence as before:

cmd:*** CONNECTED to LOADER VIA K4ABT-1,615386
Call being Setup
Call Complete to LOADER-0 @
3100912987
ROSE X.25 ROSE switch Version
9##### by Thomas A. Moulton W2VY
OK

Initially we will install only three of the .LOD files into the switch. Later as you become familiar with the ROSE SYSOP's manual, you may want to use the MEMSIZ.LOD and USERS.LOD features in the switch. For now we will use only the **INFO.LOD, HEARD.LOD,** and the **CONFIG.LOD** file/features in the switch.

Again, press and hold the ALT key while pressing the R key. This time we will highlight the INFO.LOD

file. Press <enter> and INFO.LOD will be sent to the switch LOADER. Do not touch any keys on the PC keyboard while the upload is taking place. The STATUS (STA) LED on your TNC will indicate the upload to the switch as long as the data is flowing. When the upload is complete, the STATUS LED will go out.

On the BUXTERM screen, in the right portion of the STATUS line, there should be an **OK=3** displayed. In either case, this assures us the INFO.LOD upload was successful.

Without disconnecting from the switch, continue to upload the other two files (HEARD.LOD and CONFIG.LOD) using the same method that we used to send INFO.LOD to the switch. Each time we perform the ALT R to send a file, the **OK=#** counter will reset to zero (0). This provides an accurate count as the OKs return from the switch. Three OKs are received to indicate to us that the .LOD file upload was good. After the .LOD upload is complete, the screen display should look similar to the following:

OK
OK
OK

LOADER — 615386 — BUXTERM - Version-#.#
— OK=3 -

To confirm that all three .LOD feature files are in the switch, we can send a colon and ten zeros **:0000000000** to interrogate the switch for an inventory as to which .LODs have been loaded into the LOADER. If we were successful, we should receive a record of the files, from the switch, similar to the following:

Entry #0 **LOADER-Application** Boot Interface Version 1.1
Entry #1 **INFO-ROSE** X.25 Switch Information Interface Version 2.1
Entry #2 **HEARD-ROSE** X.25 Switch HEARD Interface Version 1.2
Entry #3 **CONFIG-ROSE** X.25 Switch Configuration Interface Version 2.2
OK

We have completed the uploads to the LOADER module of the switch. Now **DISCONNECT** from LOADER!

The Final CONFIGuration

Recall that we loaded CONFIG.LOD into the LOAD-ER module of our switch. This file (CONFIG.LOD) created another module within the ROSE switch that will enable us to connect to and upload the specific table file that has been prepared specifically for this individual switch. For the purpose of this document we will upload *only* .TBL files to the CONFIG module.

Next we connect to the CONFIG module in this manner:

C CONFIG V K4ABT-1,615386 <enter>

As soon as we connect, we receive the following text from the switch:

cmd:*   **CONNECTED to CONFIG VIA**
          K4ABT-1,615386
          Call being Setup
          Call Complete to CONFIG-0 @
             3100615383
          ROSE X.25 ROSE switch Version
             9##### by Thomas A. Moulton W2VY
          OK

Notice that both the module *title* **CONFIG**, and the *address* **615386** are displayed as shown in this example:

CONFIG — 615386 — BUXTERM D Version-#.#
— OK = 3 -

The word CONFIG that follows the CONNECTED TO triggers a software "trap string" that sets up an additional safeguard to prevent the SYSOP from uploading a .TBL file into the wrong ROSE switch. As before, when we were connected to the LOADER module, use the ALT R to send the file(s) to the CONFIG module. However, this time there will be a noticeable difference: BUXTERM will prompt the SYSOP for a YES or NO answer.

When we execute the ALT R, a window will appear in the center of the screen similar to the one shown below:

```
Send 615386.TBL to the Switch?
              Y/N
```

If we press **Y**, the .TBL file for the connected switch will automatically be selected and sent to the switch.

If we press **N**, we will be switched to the directory to make another selection to send to the switch's CONFIG module. **Use extreme caution** to prevent uploading the incorrect .TBL file or the wrong version—i.e., 90713 vs 910910.

It is best to keep different ROSE switch versions of the code on separate subdirectories and/or disks (especially the USERS.LOD). Use only the support files that are supplied with the original disk version. When the 615386.TBL upload is complete, the window will display the "hex" code until the complete file is sent.

Eleven or Twelve "OKs"?

To let us know the upload is a success, the switch will send us eleven OKs. With BUXTERM.EXE there is no need to count the OKs, because BUXTERM will do the counting for you. At the time you execute the upload with the ALT R command, the OK counter is started. When the upload window clears, the OKs will appear in the normal vertical column on the screen. At the same time an OK = # will appear on the STATUS line. When a CONFIG upload is complete, the status line display will appear similar to the following display:

cmd:C **CONFIG V K4ABT-1,615386**

cmd: ***   **CONNECTED to CONFIG VIA**
          K4ABT-1,615386
          Call being Setup
          Call Complete to CONFIG-0 @
             3100615386
          ROSE X.25 ROSE switch Version 9#####
             by Thomas A. Moulton W2VY
          OK
          OK
          OK
          OK
          OK
          OK
          OK
          OK
          OK
          OK
          OK
          OK

— CONFIG — 615386 — BUXTERM — Version #.#
— OK = 11 -

Note: When using ROSE Version 901111 or above with the PASSWORD installed, an additional OK will be displayed, for a total of twelve OKs at the completion of the ######.TBL file upload.

DISCONNECT now! The "test" configuration upload is complete! Remove the power, assemble the system, and go install it at the site.

Once the new switch is installed, be sure that both the STA and CON LEDs are cycling at one second interval(s). Return to the QTH and go through the same steps as we have just discussed. Perform all the uploads again, as we described.

It will be necessary to upload all the .LOD files, since the battery-backed RAM maintains only the routing tables that are in the .TBL file. The .LOD files are *not* supported by the nonvolatile, battery-backed RAM.

I recommend including the BOOTER.LOD reset before proceeding with the overall configuration of the switch(es). Thereafter, use BOOTER.LOD only when there is reason to believe the switch is locked up, or there is a new callsign/configuration being added.

Remove The CONFIG Module To Prevent Unauthorized Access

After the configuration is complete, reconnect to the LOADER module, and as before use the interrogate command **:0000000000** to check the switch. This also permits us to identify which position the CONFIG .LOD module is in. It is important to know the correct number, or position, of the module so that we do not delete the INFO or HEARD .LOD file/features from the switch.

To confirm the position or number of the CONFIG.LOD file, we send the colon and ten zeros (:0000000000) to the switch, whereupon it gives us a display that shows the order of the .LOD files in the LOADER.

Our display will resemble the following:

Entry #0 **LOADER-Application** Boot Interface Version 3.1
Entry #1 **INFO-ROSE** X.25 Switch Information Interface Version 3.1
Entry #2 **HEARD-ROSE** X.25 Switch HEARD Interface Version 3.1

Entry #3 **CONFIG-ROSE** X.25 Switch Configuration Interface Version 3.1
OK

In this illustration notice that we loaded the CONFIG.LOD module last, which puts it into position #3. To remove the CONFIG module we will send the following command to the switch (LOADER):

:0203000000

The :02 is the DELETE command; the third and fourth digits (in this case 03) are the *position of the module,* or "file to be deleted." The balance of the ten positions should be zeros. Thus, the 03 will delete the CONFIG module. If the wrong number is used, the incorrect module or feature file could be deleted. The only way to restore it is to again upload it into the switch LOADER. A feature file (.LOD) can be uploaded to the LOADER of the switch at any time and without going through the complete configuration of the switch.

Final Notes: A hardware modification is necessary when the ROSE is used in a gateway configuration. That modification is covered in the next section of this chapter when describing the modification to the TNC when used as a node-to-node gateway. All current-production TNC-2 clones are already equipped with this mod. Do not confuse the ROSE and node gateways, as they are not presently compatible for use in a ROSE-to-node gateway application.

Our ROSE switch—building, installation, and configuration—is complete.

Definition of A Node

A short description of the various node configurations is necessary at this point. Read this material carefully and observe the illustrations so that you will be prepared to determine the best application for your needs and area.

I. NODE
    a. Usually a TNC equipped with a network EPROM which enables level 3 hierarchy combined with some level 2 features. The primary advantage of network nodes is the automatic routing capabilities.
II. GATEWAY
    a. A gateway is exactly what the term implies. The user is provided with intercommunications between different frequencies, different LANs, and different

baudrates. Multiple node/gateways can be enabled which will give access to many other nodes or gateways in a star configuration. (See figure 7-2.)

b. As a matter of fact, the gateway will perform in any of the above-mentioned configurations while acting independently as a single node. This heading will also cover the node complex and node clusters. (See figure 7-3.)

III. BACKBONE

a. The backbone is usually constructed with access from LANs along its route. If possible, the backbone should have a faster baudrate than that of the ingress LANs in order to move data more effectively.

b. The backbone frequency will bear a resemblance to the standard throughput frequency, except that it is usually on a UHF channel and provides limited access by end users. This access is only allowed at the Local Area Access Ports (LAAP) along its path. The difference between the backbone and the trunk will soon become apparent. (See figure 7-4.)

IV. TRUNK

a. Trunks differ from the backbone in two ways. Trunks are designed with point-to-point or dedicated nodes at each end. The trunk does not allow "end user" direct access.

b. Access to a trunk is possible only through routing to or from backbone nodes at the level 3 access points. The routing is never acquired via digipeaters. In many cases the routing is "fixed" by the SYSOPs, and therefore only limited access to the trunk is allowed from the "fixed" routing within the backbone nodes.

c. The firmware in the EPROM of the trunking node is developed for the exclusive purpose of restricting user access. The end user is given indirect access to a trunking link via the nodes and the backbone.

d. At this writing, I know of only one type. It is the TN11-I.EPR. Do not confuse this code with the TN11-E.EPR.

e. Many trunks which support this type of link will have a much higher baudrate than the LAN or backbone nodes.

Note: The firmware does not set priorities with regard to the amount of ingress and egress traffic.

How To Build or Upgrade Your Node Network

Now that we've learned what the nodes, gateways, backbones, and trunks are, let's get busy with the task of building them. Select the one which best suits your

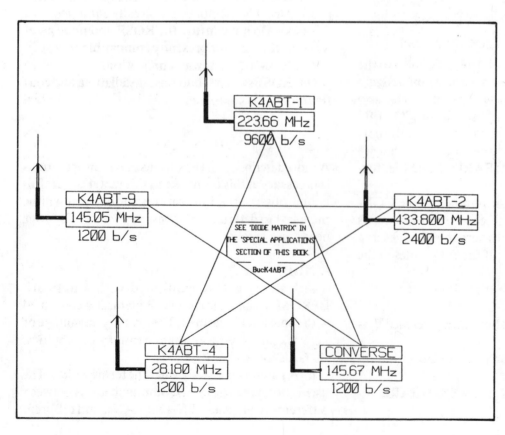

Figure 7-2. The star-shaped configuration is a cluster complex that enables easy access to nodes, backbones, trunks, gateways, and in some cases CONFERENCE nodes. Notice how the star configuration allows connects from one frequency to another.

Figure 7-3. This is another node complex that shows how a cluster of nodes may be interfaced to enable paths into many headings.

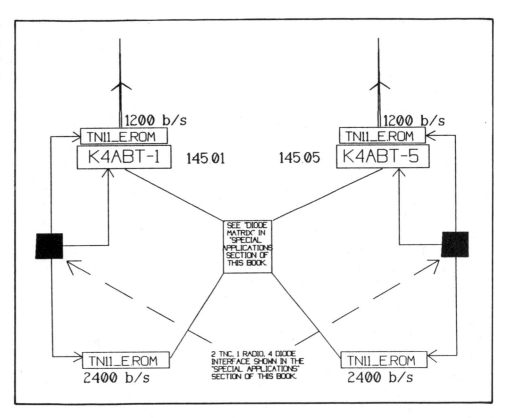

requirements or application and go to that section of this chapter. The only special requirement is the network node EPROMs. There are several sources from which you may obtain the firmware. Other SYSOPs in your area may provide more information.

The TAPR TNC-2 and clones that are of recent manufacture can be used with most of these systems. It is necessary that most of the TNCs used in these applications have the correct memory and mods that make them Net/Rom or TheNet EPROM ready. The MFJ-1270B, MFJ-1274, PacComm Tiny 2, and PacComm 200 are produced with all the current mods that make them "Network EPROM" ready.

The TNCs I used to develop this chapter are the latest production units. There is one additional modification that is required in the TNCs that are to be

Figure 7-4. The Local Area Access Port (LAAP) is LAN node K4ABT-1, while K4ABT-5 is a port of the Limited Access Backbone (LAB).

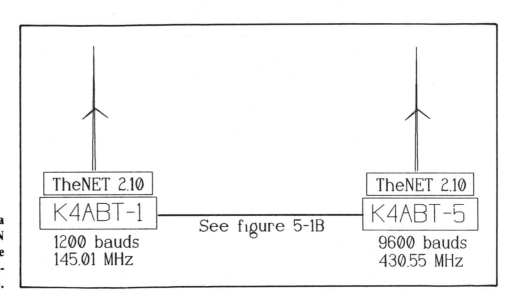

used in the node clusters, or in a node complex. So that I may prevent confusion with these systems at a later date, I make the above-mentioned modification to all the TNCs at the time they are prepared for node service. The reason is that later I may want to add more nodes or switches to the complex, whereupon it becomes necessary to add this additional jumper modification inside the TNC.

The following modification is present in the current production TNC-2 clones. This is the same modification necessary in the ROSE switch gateway we discussed earlier. The modification is minor and involves the addition of a jumper connected between the RS-232 connector, pin 23, and pin 1 of JMP 9. The jumper enables a special kind of node handshaking and flow control between nodes in a cluster or node complex. The mod also helps prevent two or more nodes in a complex from trying to activate the PTT lines to two radios operating on adjacent frequencies. This jumper *must be installed* when the TNC/node is to be used in conjunction with other nodes and gateways that communicate with a "neighbor" node via the RS-232 port. (See figure 7-5.)

The latest version of the MFJ-1270B/T has the jumper from J1, pin 23 to JMP9, pin 1 already installed. To confirm if the jumper is already installed, look to see that instead of bare holes, a 20-pin modem header is visible at J4.

Notice the jumper from pins 10 to 23 on the DB-25

connector at each end of the specially wired RS-232 cable. (See figure 7-6.)

When more than two nodes are used in a cluster, it is necessary to build the diode matrix that is shown in figure 7-7.

Each time another node is added to a cluster of nodes, the internal software commands will need to be modified to accommodate the additional nodes, gateways, or switches. We will cover these software changes later in this chapter.

Nodes, Gateways, and Backbones With 2400 bps Modems

With the sudden interest in 2400 bps PSK and the many users who are moving up to 2400 bps, there will more than likely be a need to update or modify your present nodes. You will need to provide both gateway and node capability for the LAN users who are rapidly moving up to 2400 bps. This is really easier than it sounds. The addition of one TNC equipped with 2400 bps and the cable described in figure 7-8 may be all that is required to give your system both these enhancements.

If your application requires the addition of the 2400 bps modem, see the "Packet User's Notebook" column in the May 1990 issue of *CQ* magazine. Some TNC-2 clones can now be purchased with the 2400 bps modems already installed.

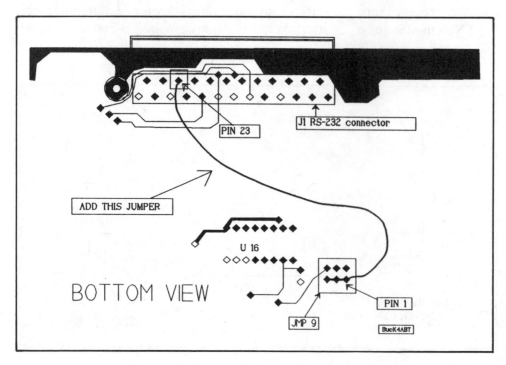

Figure 7-5. Placing a jumper between RS-232 connector pin 23 and pin 1 of JMP 9 enables handshaking when two or more TNCs are interconnected.

Figure 7-6. This interconnect cable differs slightly from the cable used with the ROSE switch. Notice the jumpers from pin 10 to pin 23 of each connector.

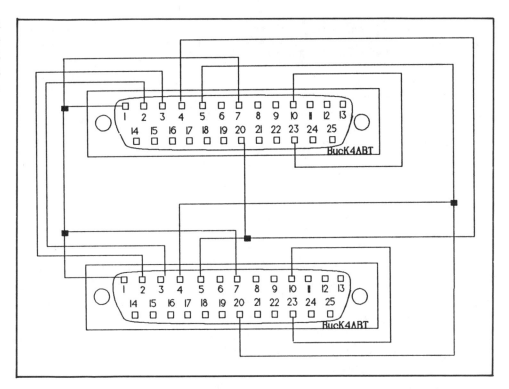

An Illustration Is Worth 10,000 Bytes

For the benefit of those new SYSOPs who are about to venture into the world of node, gateway, and backbone construction as a result of this chapter, you will be pleased to discover that I have drawn most of the illustrations so that you can feel comfortable when building your system just by looking at the drawings. I drew the illustrations with much of the information contained in the graphics. This way you will have a better understanding of the node parameter configuration that we are about to review.

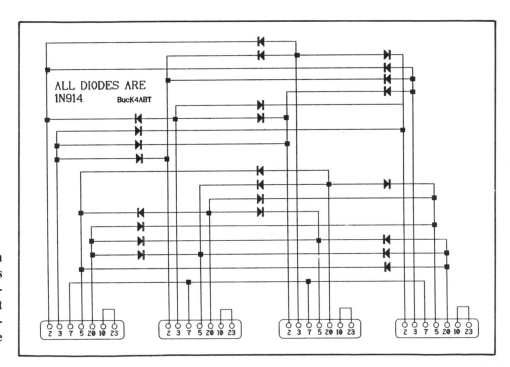

Figure 7-7. When more than two nodes are interfaced, it is necessary to construct the diode matrix shown. A slight variation of this matrix is employed when interfacing more than two ROSE switches.

The MFJ-1274 is a packet data controller equipped with 2400 baud modem. (Photo courtesy MFJ Enterprises, Inc.)

Setting The Parameters

This part of the node, gateway, and backbone building is that portion of the system which gets the most attention and is the least understood. I speak with the utmost authority on this very touchy subject. Even the writers of the firmware code are at odds as to how the parameters should finally be set. Any configuration I include herein may be changed to suit your needs, applications, or your own personal wishes. I am *not* about to impose my node configuration tables on any newcomer, and I am even less likely to try them on the old timers, who jump sky high each time I make a parameter change to a node in our LAN.

The parameters supplied with the network code you decide to use will be a guide with which to begin. I've also discovered that SYSOPs who make changes to other parameters will sooner or later return to the original setup.

Packeteers, Start Your Engines

After you've made the necessary hardware modifications, connect the TNCs to a terminal and turn on the power. Be sure you get the correct "Network Rom" sign-on message. Verify the callsign of each node, as each node should have a different SSID embedded within the sign-on message. Confirm that the callsign is correct.

Connect to the node using the **ESC-C** command and < enter >. The node should respond with "CONNECTED TO," etc. Enter the node identifier using the **IDENT** or **INFO** command. Carefully follow the instructions in the network node manual.

You may wish to use the local airport identifiers or you may do as I have, and use a portion of your callsign—e.g., TNC A is "ABT1" and TNC B is "ABT5." Other than the obvious, there are several reasons to use separate callsigns on the nodes. Once you are connected, configured, and have the correct callsigns installed, you should be able to "gateway" between the two nodes.

Finally, check the password by executing the **ESC-P** command. The password string can contain up to 80 characters. There should not be any spaces, line feeds,

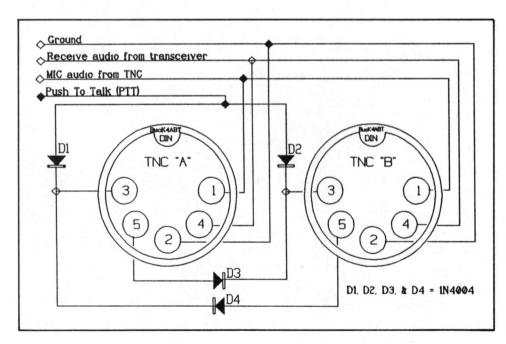

Figure 7-8. This special "radio port" interface cable is useful when adding a 2400 baud node to an existing 1200 baud node or ROSE switch. This configuration helps prevent both switches keying at the same time. To add gateway capability, include the interface cable shown in figure 7-6.

or carriage returns. The longer the password, the better, but stay within 80 characters.

Completing The Node

Do an **ESC-D** to disconnect. Then **ESC-Y-0** (zero) will disable the host connections. Finally, remove the terminal. I would recommend operating the system in a "test" environment for a few days, or until you are happy with the behavior and operation of your setup.

NOTES

Chapter 8

Antennas and Digital RF Communications

This chapter deals with the specifics related to various types of antennas. We will consider everything from an isotropic radiator (dipole) to a rhombic. Putting it another way, we will research antennas from the customary to the complex. I'm not suggesting that any digital radio user should select one antenna over another. Use whatever you have, or choose the antenna that best favors your needs and environment.

Packet radio is one of those communications modes that will tell on the system operator if he or she fails to provide the antenna that has the best radiating and capture effect to it. In fact, if the antenna is not constructed and erected so it will provide good capture to signals and have the lowest noise component with respect to terrestrial noise, then no one is to blame except the operator in charge of the installation.

I am as meticulous as A.J. the day before a race. My advice is don't just walk the race track; look for the bumps and crevices. The antenna for your packet station is about to become your doorway to the world. Everyone who has spent any time around me will affirm that I won't skimp when it comes to my antenna.

I am very particular where my antennas are concerned. When I go to buy cable or connectors, I specify silver-flashed connectors and cable of the best quality. That is the part of my station that will get the least attention after it is installed, so I want it to withstand the elements and provide dependable communications for a long time. I am very picky about the antennas and associated components of my antenna system. With over 40 years as an amateur radio operator and

over 25 years as chief engineer for a group of radio and television stations, I learned a very valuable lesson early on. Signal quality begins at the tip of the antenna, and it travels down through the transmission line

A six-element VHF Bandmaster quad is shown here overlooking an Alabama sunset. (Photo courtesy Alabama Amateur Electronics [AAE]/Bandmaster Quads.)

and reflects off the operator at the other end. Let your reflection be a good one.

Radiation and Resonance

If we were to feed an RF signal to a piece of wire suspended in the air, the signal would radiate over a wide area. To obtain maximum coverage, the wire should be a resonant length at the transmitter frequency.

Antennas can be constructed to radiate with directional, omni-directional, and bi-directional patterns. The kind of pattern desired depends on the coverage-area requirements. Likewise, the type of antenna selected will determine the kind of pattern you will have. Another major factor in antenna selection and installation is the distance above the ground at which an antenna is suspended. Antenna theory as related to antennas suspended in free space states simply that the ground below will provide a reflection, or mirror effect. This mirror effect gives an antenna the appearance of having greater gain when the antenna is mounted at distances that are ''in phase,'' or a given wavelength above the earth. The greater the height, the greater the gain.

It is understood that radio waves travel at the velocity of light in free space. Therefore, radio waves travel at 300,000,000 meters per second, or close to 186,000 miles per second. A formula for determining the resonant length of an antenna for a given frequency is based on the speed of light theory. One of the first formulas that we learned when studying for an FCC license was the formula for computing the length of an antenna. The formula for the length of an antenna, expressed in meters, is:

$$\text{Wavelength (meters)} = \frac{300{,}000{,}000}{\text{Frequency (Hz)}}$$

Band Considerations

Up to this point we have discussed some simple, but general antenna theory. Now we go directly to the application of antennas and the frequency of operation. This is another way of looking at our needs and requirements.

Almost all antenna basics can be analyzed in terms of the elementary dipole. A dipole consists of two charges of opposite polarity. In the case of a real antenna, the charges take the form of two elements which receive signals of opposite polarity. (See figure 8-1.) We have learned through theoretical as well as practical experience that any antenna will exhibit the same characteristics whether it is used to receive or

to transmit signals, provided the impedance of the feed system and the radiating elements are the same.

Several kinds of high-frequency (HF) antennas are simple in design and construction. The simplest is the end-fed or long-wire antenna. In most cases the long-wire is a one-half wavelength antenna cut for the frequency of operation. When designing or building an antenna, the size or diameter of the element (wire) should be taken into account. A well-known formula for determining an antenna of one-half wavelength, expressed in feet, is:

$$\text{Length (ft.)} = \frac{492}{\text{f (MHz)}}$$

Ground Influence

Another consideration to be added to the computations is the influence of the ground on the antenna system. It is stated as follows:

$$\text{Length (ft.)} = \frac{468}{\text{f (MHz)}}$$

This ground influence is often referred to as the K factor. The K factor is a constant of 0.95 that is used to make the formula follow a more accurate standard with wire sizes close to number 16 gauge. For our purpose, the above formula is most accurate for determining antenna length between 1.8 and 30 MHz.

As we leave the HF spectrum and enter the VHF and UHF spectrum, the antenna design changes, too. Most antennas above 30 MHz are constructed of aluminum tubing or rod, and since the tubing can sometimes be much larger than the wire size mentioned earlier, it stands to reason that a new K factor must be considered. An antenna handbook will usually have a listing of K factors for different diameter-to-length ratios.

At HF frequencies I use everything from a doublet to a beam for my packet operation. If I want to get on the air quickly, then I go for an old tried and proven antenna called the *dipole*. (See figure 8-2.) The dipole by nature is a center-fed antenna. If we consider all the different factors related to this antenna, we will find that it has the closest impedance to the coax feed lines that are available to us today.

Theory says that the impedance of a wire antenna measured at the center is approximately 72 ohms when it is at a height of about one-half wavelength above ground. The dipole is usually suspended between two poles or trees and supported at each end by a nonconductive material (insulator).

Figure 8-1. Shown here is an illustration of the basic dipole, which consists of two elements or charges that receive signals of opposite polarities.

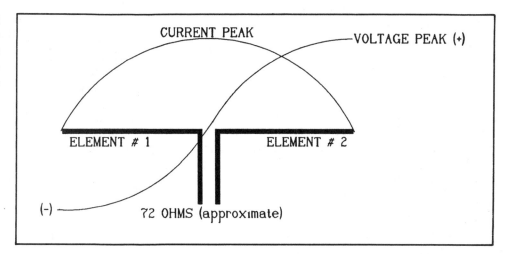

Even as a coil and capacitor form a resonant circuit, so does an antenna. Depending on the ratio of inductance (coil) to capacitance, or L to C, these two have something in common. An antenna has Q, as does the coil and capacitor. This Q affects the gain versus bandwidth product of the circuit, whether it is coil/capacitor or the antenna.

Whereas the Q is affected by the ratio of L to C, the Q of the antenna is affected by the size of the wire used in its construction. If the antenna is a VHF or UHF antenna, the Q is more pronounced as the size of the tubing is increased. A lower Q will provide increased bandwidth. However, there is a trade-off at this point, since we don't get something for nothing. We decrease the gain factor as we increase the bandwidth factor. Conversely, if we decrease the size of the tubing, the Q will increase; thus, the gain also increases, but the bandwidth decreases.

Horizontal Versus Vertical Polarization
Short and to the point, a vertically polarized antenna can be good when used in a beam configuration at VHF to reduce absorption of atmospheric noise.

It requires that reflector and parasitic elements be added to improve directivity and increase gain. (See figure 8-3.) On the other hand, if only a vertical driven element is used in a ground-plane design, then the resulting pattern, when viewed from above, will appear as if it were a doughnut with a small hole.

It goes without saying (but I will anyway), vertically polarized HF beam antennas with long elements, and towers with guy wires affixed near the top, do not work well together mechanically. Something will have to bend or break—either the guy wire or the element, since they are in each other's path. For this and other reasons most HF beams are horizontally polarized.

Voice Versus Digital
Don't be deceived by the heading of this section. I am not about to begin an argument over these two modes. My intention is to look into the types of antennas that are best suited to the digital mode of communications, as related to the antenna commonly used for voice communications. From the beginning of this chapter we have moved in this direction.

If it is distance you want, then the class of beam

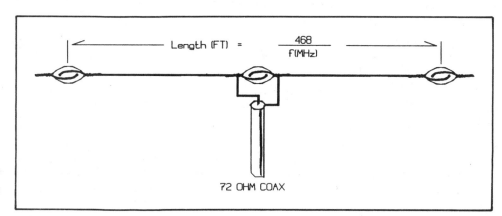

Figure 8-2. To get on the air fast, use this tried and proven antenna—the dipole.

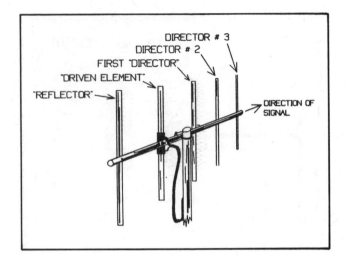

Figure 8-3. A vertically polarized antenna is a good selection due to the manner in which it reduces absorption of atmospheric noise. (Note: The reflector is the longest element.)

antenna we use for voice will be sufficient. If it is coverage you prefer, again I prefer the beam-type antenna as a power booster. I tend to try for a happy medium with respect to the digital and/or packet modes. The Yagi-type antenna in a horizontal configuration is one way to go if you want coverage and reduced wind resistance.

The happy medium I spoke of above takes the form of a vertically polarized Yagi or a cubical quad. The reason I chose the latter is because the quad is well known for its favorable gain/bandwidth characteristics. Second, the quad offers a better signal-to-noise ratio because *influence from terrestrial noise is greatly reduced* when receiving with a cubical quad antenna. This inherent rejection to terrestrial noise is one of the reasons we might consider the quad for use in a digital data medium. (See figure 8-4.)

Field Intensity Measurements (VHF/UHF)

If you would like to know how well your packet station or node is performing, you can use relative field intensity measurements to determine the results. Following is a system that I use.

Put the packet station on an unused frequency for your area, and set the beacon time so that it will identify every 5 minutes. Add some extra TXDelay so that the beacon will have a little longer power-on time than it normally would. Use low power if you have that facility, and begin the measurement.

Pinpoint and number several (minimum of ten) locations on a map of your locality. Maintain as much circularity to your route and locations as possible.

Be sure your mobile unit (1) will receive the frequency of the beacon, and (2) has a dependable S-meter.

Use an omni-directional mobile antenna, preferably a magnetic roof-top antenna. Drive as near as possible to the locations you pinpointed and park until you receive one of your beacons. Record the relative signal strength (S-meter reading) and the location number, and continue to the next location for another measurement.

By making these measurements you will arrive at a relative field pattern for your antenna and a presentation of its performance. When you plot the curve of the field intensity measurements, you can quickly observe any "holes" (nulls) in the signal as well as any unusual pattern anomaly.

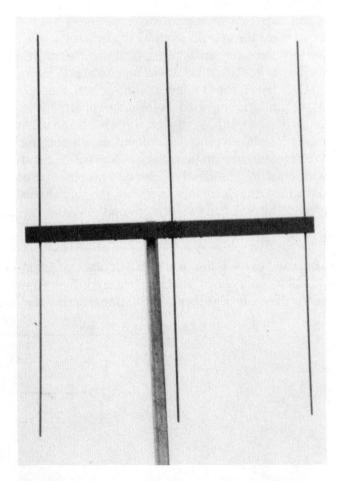

This MFJ three-element Yagi is normally mounted vertically polarized for packet use. (Photo courtesy MFJ Enterprises, Inc.)

Figure 8-4. The quad has an inherent rejection of terrestrial noise. This makes the quad a better antenna to use at VHF with the digital modes.

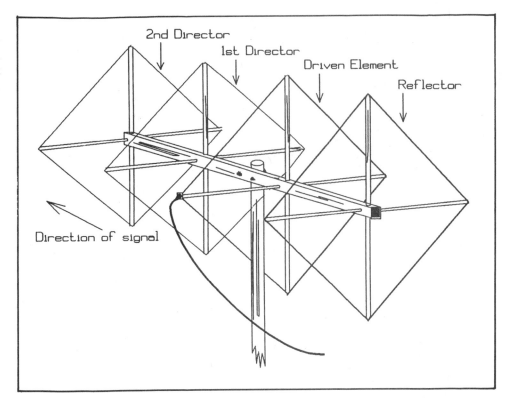

When you have completed the field measurements, turn off the beacon and return the TXDelay to its normal setting.

Point-To-Point Communications

There are fewer repeaters (digipeaters) for VHF packet than there are for VHF voice operations. This condition alone makes it favorable for the packeteer to choose a beam or other type of directional antenna. This way you can reach those far-off nodes that are just out of range for your omni-directional antenna.

Even though packet is a "store and forward" medium, the ardent packeteer and digital amateur should choose an antenna wisely, because packet is a point-to-point medium that is based on simplex operations. "The most reliable path is one which can be regarded as a 99% path by the user, since the perfect path is considered less than 100%." If you find you have reached a point where your VHF signal is marginal, and you would like to make it a "95%" path between two beam headings, maybe you should try stacking another antenna of the type you are using. If your beam is a Yagi, you can achieve another 3 dB of gain by stacking a similar Yagi one-half wavelength away. (See figure 8-5.)

Shown here is a Cushcraft 124WB four-element, wideband 2 meter vertically polarized antenna. (Photo courtesy Cushcraft Corp.)

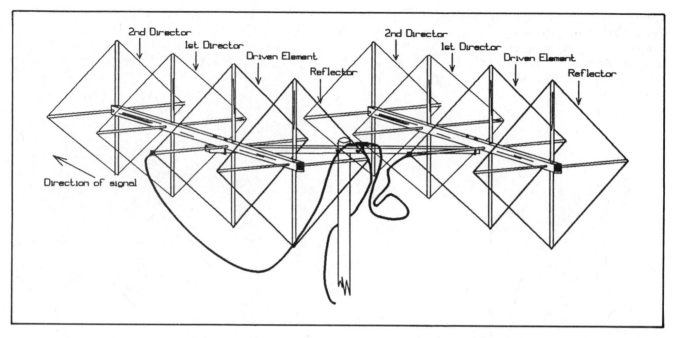

Figure 8-5. The same rule applies when stacking quads: A 3 dB improvement can make the difference between making the connection or the loss of a packet contact.

The Thoroughbred

As with any hobby, the thoroughbred does not come cheap. Specifically, the *rhombic* antenna is the best of all worlds, but it requires plenty of running space. A rhombic is a four-sided antenna with two of the sides slightly compressed towards the center, forming a diamond shape. (See figure 8-6.) Each one of the four sides of a rhombic will have in excess of ten wavelengths. The feed point shown in the illustration is fed with 450 to 750 ohm open-wire feeder. The end

of the rhombic that is opposite the feed point is in the direction that the rhombic is pointed. This point is

Figure 8-6. Compressed slightly from the sides, the rhombic is diamond shaped.

The ground-plane antenna is used in many ROSE switch mountaintop installations. (Photo courtesy MFJ Enterprises, Inc.)

These are stacked VHF and UHF Bandmaster quads inside a 10 meter quad. (Photo courtesy Alabama Amateur Electronics [AAE]/Bandmaster Quads.)

normally terminated with a resistor of 500 to 800 ohms.

The rhombic is not a movable antenna, and it does need many wavelengths per side to be effective. Twenty wavelengths per side could render forward gains in excess of 15 dB at VHF.

The antenna for digital communication, as well as any other modes of communication, is only as good as the transmission line that feeds the antenna. Consult the handbooks and the catalogs for the latest and greatest coax or transmission feedline. Look at the manufacturer's printed specifications for a given type of feedline or coax. The main points of interest are the specs regarding the loss factor (expressed in dB) per hundred feet, the velocity factor, and the frequency at which the measurement was taken. Over the long haul the "hard-lines" or multiple-shielded coax cables will prove to be the better value.

Assuming that your next question is "Why are we interested in the velocity factor of the feedline or coax?" read on.

The coaxial (coax) cable or the transmission line plays a major role in antenna performance. The coax is a very vital part of the overall antenna system. It has a personality of its own and can wreak havoc if it is not cut to or "tuned" for optimum performance along with the antenna. It is even more important to note that antenna performance will depend on the behavior of the transmission line at the time of antenna tuning or setup. In other words, if the coax is not prepared before the antenna is tuned, then tuning of the antenna will not render optimum performance.

The coax is the "life-line" that delivers the energy to the antenna. Since the energy is handled by the coax, this means the coax is either an external extension of the "tank circuit" or it is part of the antenna. But which is true? This is not a trick question, but a way to make a statement that can easily be remembered. The antenna feed line is *both,* because the complete antenna system is part of the tank circuit.

Now we are beginning to understand why the antenna should be tuned. This is where we discover that tuning the antenna is another way of achieving "maximum transfer of energy."

To better understand antennas and their feed lines, let's consider this analogy: If we imagine a pump with a capacity to deliver 100 gallons (watts) of water/pressure per second, this would mean that a pipe of ample size would be needed to deliver the water to the distribution tank. This pipe should be the exact size to correctly handle the pressure. If the pipe were too small, a "back pressure" would occur. If the pipe were too large, a loss of pressure would occur. The "back pressure" could represent standing waves, and the loss of pressure could represent a mismatch of impedance at the antenna. As I stated, the analogy is crude, but it helps us to understand why the coax (pipe) size and length are important.

The Velocity Factor Must Be Considered

Now we understand why tuning the coax should be the first item we consider. But how do we tune a piece

Type	Impedance	Velocity of Propagation		MHz	Loss/100 ft.
RG-8U	50 ohm	Solid Polyethylene	66%	100	2.2 dB
RG-8U	50 ohm	Foam Polyethylene	78%	100	1.8 dB
9913	50 ohm	Semi-Solid Polyethylene	84%	100	1.4 dB

Table 8-1. Some often-used coaxial cables and their specifications. More complete information is available from most cable manufacturers.

of coax? A feed line that is measured and cut in exact multiples of one-half (electrical) wavelength at the operating frequency will display the same impedance at each end when connected to an antenna that is cut and tuned to a given frequency.

There is one small caveat, and that is the "electrical" wavelength. To determine the "electrical" one-half wavelength for our operating frequency, we must know the "velocity factor" of the coax that we plan to use. Table 8-1 is a short list of some often used coaxial cables and their specifications. These specs include the velocity factors as stated from the manufacturers' printed specs. However, the specifications are not complete. Further information is available from most coaxial cable manufacturers' master catalogs.

By using a modified version of the one-half wave antenna formula we discussed earlier we can now determine the required length of our coax in multiples of a one-half wavelength from this formula.

$$L \text{ (ft.)} = \frac{492 \text{ V}}{f \text{ (MHz)}}$$

where V is the velocity factor, L is the length in feet, and f is frequency in MHz.

Let's assume that we have assembled our antenna and adjusted it for 145.050 MHz. We plan to feed the antenna with RG-8U polyethylene (foam) cable. We find the velocity factor of the RG-8U (foam) coaxial cable to be 78%. Our goal is to determine the length

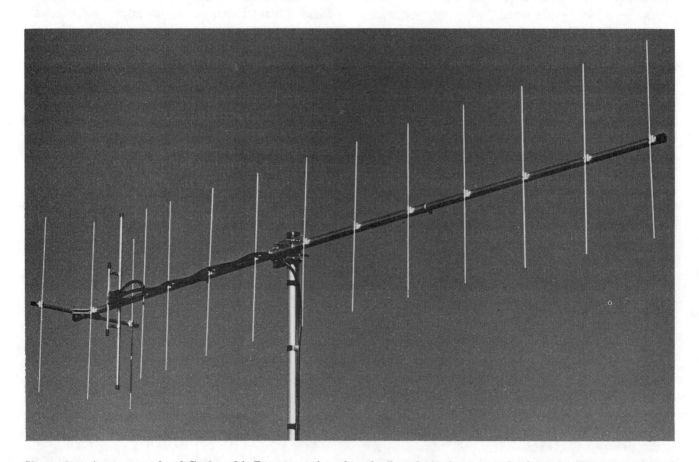

Shown here is an example of Cushcraft's Boomer series of vertically polarized antennas for 2 meters. (Photo courtesy Cushcraft Corp.)

(L) in feet of the coax feedline, but first we need to know the length of one-half wavelength of the RG-8U cable for the frequency of 145.050. We calculate in the following manner:

$$\text{Length (ft.)} = \frac{492 \times .78}{145.050 \text{ MHz}} = \frac{383.76}{145.050} = 2.64 \text{ ft.}$$

If we care to go a step further, we can convert the feet to inches and arrive at a half wavelength of RG-8U foam coax for 145.050 MHz. The length is approximately 31¾ inches. The distance to our antenna is about 48 feet. To be on the safe side, we want the cable to be the next one-half wave longer than the distance to the antenna. By dividing the 2.64 feet of the one-half wave of cable into the 48 foot length of cable we need to feed our antenna, we arrive at 18.18 times the 2.64 feet. Let's go to the nearest greater length and call it 19 times.

$$19 \times 2.64 \text{ ft.} = 50 \text{ ft. 2 in. (approx.)}$$

A Shortcut

The easy way to make this measurement is to use an antenna bridge. There are several of these units around, but I used one that is designed for use between 1.8 and 30 MHz. It is manufactured by MFJ and the model number is MFJ-204B. If you have one of these antenna bridges, then follow procedure number II in the instruction manual. By using the antenna bridge method, there is no guesswork involved, and the results are sooner realized.

The Antenna Is The Key "Element"

Some of our explanations may seem a bit long, but the final measure of the radiating elements of the

The MFJ-204B antenna bridge is designed for use between 1.8 and 30 MHz. (Photo courtesy MFJ Enterprises, Inc.)

packet station will manifest itself in the performance when it is put on the air. More than once I've responded to a call for help from a fellow packeteer who tried to make do with a piece of wire and a random length of feed line. Only after rebuilding the antenna system as I've just described will the packeteer really understand the full importance of the antenna system to a packet station. Consider the radiating element of the packet station as if it were a long-term investment, because it surely is!

NOTES

Chapter 9

Packet Radio RFI Causes and Cures

$gnd < \frac{1}{8}\lambda$

In the hobby of packet radio there is no place for RFI. Some forms of amateur radio can tolerate some effects of RFI, but packet radio seems to have its own type of rejection to the presence of RFI. The nature of digital communications is in the rise and fall times of some types of digital data. Other modulation techniques besides FSK are now being used to increase the speeds of digital data through the airwaves. Soon we will see the results as the base-rectified RFI begins to appear with some of the current digital modulating methods such as Minimum Shift Keying (MSK) and Quadrature Phase Shift Keying (QPSK).

Let's explore the reasons why your packets may not be printing on other packet stations' screens, or the incoming packets may not be printing on your screen. The reasons may be the fault of something other than incorrect terminal parameters.

Hum Versus RFI

It could very well be that you have hum and even distorted audio caused by poor power supply regulation or radio frequency interference (RFI). Let's not belabor the reasons why RFI occurs, but instead let's delve into the mechanism by which we can abolish this troublesome problem.

A good ground is "worth its salt" in the HF and VHF packet station, although the ground is not the absolute answer. Use the ground as you need to, but remember this: "A ground loses its effectiveness if the length of the ground wire exceeds an eighth of a wavelength at the frequency of operation." If you happen to be using the 10 meter band, this works out

to be just over 4 feet in length. Beyond that, the ground-wire length becomes a radiator. In effect, a ground loop is created, and the problems related to RFI commence. As we go up in frequency to the VHF spectrum, we see that the wavelength gets shorter, as does the effective length of our grounds. At VHF we must be extremely careful with the ground-wire lengths (distance between the driven ground rod and the equipment), because a 20 inch length of ground wire can become a radiator or even an "absorption pick-up loop" at our favorite packet frequency. To circumvent the problem in the HF packet frequencies, I've used an artificial ground similar to the MFJ-931. This removes the worry of having a long ground lead from the cold-water pipe or ground rod to the equipment.

The artificial ground enables me to tune the reactance out of the long ground wire, and makes the ground rod appear to be very close to the HF equipment. The purpose of the artificial ground is to displace the stray RF to ground or present a quick DC path to any RF that may appear on the ground wire connected to the equipment. Thus, the RF is displaced before it has a chance to get to the system ground.

RFI at VHF Frequencies Displays A Split Personality

As technology advances, things change for the better —except this age-old enigma called RFI. There have been numerous times when I've chased the ugly vermin around my computer room, only to discover the

RFI felon would appear somewhere else. The last time I discovered noise and hum on my signal (both transmit and receive) I decided then and there to get rid of this dreaded nuisance once and for all. It didn't take long to learn that a good ground was merely part of the cure. It took time, but at length I prevailed, or at least I've had a contented packet life in recent months.

One of my encounters with RFI was in the early days of VHF packet. We were inaugurating a new venture, and as many of us recall, the TNC (sometimes called the "PAD") was not inside a metal box, nor was there a lot of circuit board trace bypassing as it is today. At best it was a couple of homebrew printed circuit boards sitting on two strips of wood. If I had the power amplifier on, when I pressed the enter key to ship a packet to a connected station all chaos broke loose. Having had a couple of connected stations tell me they couldn't print all of my packets because there seemed to be a roughness to my signal, or that I had hum on the carrier, I began looking for the source of the annoyance.

Note: An indicator of possible RFI problems is a large number of "TRIES" even though the signal strength of the two direct-connected stations is good and the modulation is optimum!

As many of us have done, I misread the description of the symptom that was given to me by the connected station, and began the search for my problem by looking at the ripple from my power supply.

Whether or not the power is a clean power source can be seen with a scope. In the absence of a scope, use a digital volt/ohmmeter and watch the I/R drop during the transition from the load to no load (transmit to receive) condition. If the voltage is no more than a few millivolt (a hundred or so, but no more than 200 mv) change in the meter movement, let it pass, and go on to some other possible cause of the trouble. RFI gives no clue as to where it will appear, whether in the transmit or the receive portion of the packet station.

If the reports indicate the presence of RFI in the transmit audio, then head for the TNC end of the interface cable that connects the TNC to the transceiver.

The first thing I do to stop RFI from becoming a base rectified DC component within the transceiver or the TNC is place ferrite beads on the AFSK mic audio lead at the TNC. The Amidon FB 73-1801 works well in this application. Another means of subduing RFI is by placing a .001 uFd 25 volt capacitor from mic input to ground. I avoid the latter step if I can because some of the audio frequencies are also bypassed to ground. In addition, most TNC manufacturers now install the correct bypass capacitors inside the TNC. To add extra bypassing external to the TNC could radically change the audio equalization and compound the problem instead of improve on it.

Many times the receive line is fed to the same connector, and while you're inside the connector shell, you may as well protect the receive lines from RFI as much as you can. Add the same type ferrite bead to the receive line as was used on the transmit audio line. You can even go a bit further and place one on the PTT wire, which in 85% of the cases will be inside this shell. The inclusion of the PTT line is optional and space dependent.

When you have completed these additions, you may want to add a few ounces of prevention, or as the case may be, "some more cure." There are some devices made of a ferrite compound that snap together to form a closed loop around the cable. They are constructed in such a manner as to be wrapped with a wire or cable, and both pieces are held together by a plastic retainer.

These chokes are quick and easy to install on any cable up to a half inch in diameter. There is no cutting or soldering to be done. Just follow the easy installation instructions and you can almost see the problems go away. In fact, if you are using one of the home computers as a terminal (e.g., the TRS-80 Color Computer or the Commodore C-64), you are already aware of the kind of RFI which causes the ripple and other picture distortion in your TV/monitors. We can use the snap-together RF choke to eliminate most, if not all, of the RFI in the video by winding several turns of the computer-to-TV interface cable through the snap-together choke. It may take more than one of these snap-together ferrite chokes to make the job complete, but the end result was well worth the effort for me. As it turns out, they must have known what we needed, because in the final clean-up I had used one on the radio-to-TNC cable, one on the terminal-to-TNC/RS-232 cable, and two on the CoCo-to-TV interface cable.

If your station has a problem with RF in the telephone lines, try placing a .1 mFd, 600 volt capacitor directly across the telephone line. In a few instances I have added a 2.5 mH RF choke in series with each side if the phone line. The RF choke should have wire size suitable to sustain the current contained in the line when ringing current is present.

The MOV

Finally we get around to the metal oxide varistor (MOV). This device can cover a multitude of crimes, some of which are in the RFI category. First and foremost is the ability of the MOV to chop noise and impulse-induced spikes from the power lines. Metal oxide varistors have been used in my computer room for many years, and I can attest to their usefulness as a protective device.

The MOV does more than reduce spikes and line noise. Its intended use is to shunt the line in the event of a power surge and pop the fuse, thus protecting the equipment up ahead. More on that later in this chapter. They do work!

The response time is measured in nanoseconds, and sometimes they don't recover. If they don't recover or they are shattered, so goes the way that your equipment could have gone. The price is minimal by comparison. The science and theory of these little devices makes for some good reading, but I won't go into it here. They can be purchased in the form of an AC plug adapter or as an add-on unit which appears very much like a large ceramic capacitor.

Experience Is The Best Teacher!

I was in my former office atop the American Center in Nashville, Tennessee. There was a storm moving in, bringing a sudden blast of cold air to our area. It was predicted earlier in the day that the temperature would drop almost 40 degrees in 24 hours. The prediction should have been a warning to me.

I looked up from the computer, where I was compiling data. I glanced out the window and noticed the sky had darkened with heavy clouds. We were high enough that the building level where I was located was surrounded by the storm. The phone rang. It was my wife, Jean Ann, WB4EDZ. She told me she had just returned home from the city to find that all the ceiling fans were not working. She also said other appliances were not working, including the garage-door opener.

Oh, no! It couldn't have been. I made a fast exit and drove to my home at the time, about 25 miles north of Nashville. When I turned into the driveway the first thing I noticed was that my Diamond F-23A was missing. So was my WinTenna 9209. I drove around to the rear of the house and looked closer. The F-23A was indeed *gone!* The 9209 was there, but the fiberglass portion of it was shredded and was barely able to support the weight of the new stainless-steel tip.

Not wanting to think the worst, I went inside. Jean Ann told me she had already checked the large-screen TV and it was okay. I felt a little relief, as that 55 inch screen TV was my birthday gift from her.

Then she hit me with the big one: "There is a bad odor in your computer room." She muttered something else about the air-conditioner not working, but the air-conditioning had just moved to last place in my mind.

My next stop was at the door of my computer room. I stopped! There was the smell of burned phenolic, resistors, integrated-circuits, and whatever other scorched electronic component smell a seasoned amateur can relate to.

I walked over to my chair, sat down, and began looking over the mess. There were signs of smoke and soot all over the room. The ends of connectors were blown out of the sockets. The one part of my station that I felt might be immune to this kind of mutilation was in a shambles. The batteries that are (were) part of my solar system were intact, but the cables that had connected them to the rest of the equipment were separated as if they had been ripped apart at the fuse holders by some mighty power. As a matter of fact, they had. The same applied to the power cables to the DC supplies that had equipment attached. And here is the real kicker: The supplies were *not* turned on!

I began checking everything that was connected to any kind of power source. Everything in the room was destroyed, trashed, or would not show any signs of even a puff of smoke when attached to their respective power sources. I opened the KAM. It was covered with soot, and the PC board was blackened. The same thing with my MFJ-1278 all-mode controller. The Alinco DataRadio 1200's were zapped as well.

The phones were all dead. Thus, we were out of the wireline telephone service for over three days. Thanks to my cellular phone we were still in contact with the rest of the world. The cellular phone was worth its weight in platinum.

The HF transceiver was sitting atop of one of my 486 PCs. When I attempted to lift it off the PC, it would not move. It was welded to the case of the PC. It was not until the insurance adjusters arrived the following day that it was removed.

Caveats, Cautions, and Concerns

I learned a lot from this loss, and we are fortunate when we consider that we were not physically hurt. There is a lot to consider when you begin to recover from such a mishap. Soon thereafter we put in several ground rods around the house, and they were bonded together to provide a better ground than the poor ground that was provided by the home-builder and electrical contractor.

I then put coax switches on all the antenna cables that came into the computer room. The purpose of these coax switches was so that I could place them into the GROUND position before leaving the room. The MFJ coax switches provide a connection to an outside ground supplied by the user.

I put a separate disconnect on the utility power entering the room. In addition, I installed a similar disconnect between my solar electric panel array and the solar storage system.

Once More, The MOV

I discovered something that at first seemed to be a coincidence until I began to realize there was a trend. I had used several MOVs (metal oxide varistors) on several appliances throughout our house. My 55 inch screen TV was attached to the utility power via one of these MOVs, and it was *not* damaged. The refrigerator was plugged into one of the MOVs, and it was not damaged. The older 8088 PC in the room where most of the damage had occurred was not harmed; it was attached to the AC power via an MOV. In all, I had six MOVs in use, and in every case there was no damage to the device that was attached to it. Needless to say, every outlet in my home was then outfitted with an MOV.

Making Some Notes

• Our VCR was connected to the same power outlet as the big-screen TV. The big-screen TV had an MOV in-line and it was *unharmed*. The VCR had *no* MOV in-line and it was trashed!

• If you have more than one computer, or if your coverage of computer-related equipment is beyond the coverage listed in your home-owner's policy, it might be to your advantage to review the policy with your insurance agent.

• List all the computer equipment on a separate rider. Include printers, monitors, and modems. Any item associated with data collection or processing is considered to be part of the computer system by many insurers.

• Amateur radios and transceivers are *not* part of the computer system, although the radio can be attached to the modem/TNC. Yes, a TNC can be considered a modem or part of the data-gathering/processing equipment by some insurance companies.

For the benefit of those who wish to prepare for a blast such as I experienced, I have listed several types and kinds of MOVs that I am now using or have used in the past. The reason I also say "have used" is because after an MOV has been hit or has done its job, so to speak, it must be replaced.

The price of the MOV is cheap when compared to the cost of the device or apparatus that it can save.

I am *not* recommending the so-called protective surge-suppressor power strips. The reason I mention these is because one of the 386 machines was plugged into such a device. Not only was the 386 trashed, but the switch in this surge-suppressing outlet bar was welded *on* by the hit. Plugged into the same AC outlet (duplex) was the old 8088 PC with an MOV attached, and it is alive and well.

Here is a partial list the MOV devices I have used: Archer (Radio Shack) No. 61-2792 w/status indicator light; Tandy (Radio Shack) No. 26-1395A twin outlet w/reset and status indicator light; and Tii transient voltage surge suppressor Model No. 428.

Another note of interest is the three latter devices have status lights which extinguish if the internal MOVs are destroyed or otherwise fail. My cellular phone battery charger was plugged into the Tii protector and survived; however, the Tii protector did not. The status "glo-lamp" no longer lights. In substance, if the light is on, the protector is okay, and if the light is out, there is no longer any protection from the spike suppressor.

The latter can be installed inside the equipment and represents a small price to pay for the amount of insurance it provides. The MOV can be found at most electronics supply stores and in some hardware stores.

With a few components and a little time you can have a better packet signal on the air and better protection for your valuable equipment. This will make your hobby more enjoyable for you and for the connected station.

Chapter 10

The Packet Bulletin Board Service (PBBS)

There are several BBS system formats, but in an effort not to confuse the reader and user, I will try to cover them in a general manner rather than build this tutorial around the specific features that are common to each individual BBS format. It is not often that we find the kind of unified cooperation shown by the BBS writers and authors in an effort to set a "standard."

Let's connect to the WB4RHO BBS:

C WB4RHO

ABY1:WB4RHO-1} Connected to WB4RHO
[BBS-PROM 1.0 BR$]

Hello New User; Welcome to the WB4RHO BBS in Headland, Alabama

First of all, the BBS doesn't know who I am, so it calls me "New User" and welcomes me to the BBS.

This is because I have never connected to this BBS before. As soon as I answer the next group of questions, I will be welcomed as "Buck," and the additional questionnaire will no longer prompt me for information. Following is the set of questions which will be sent to me on the initial connect to the BBS.

N xxxx—Enter your first name into user data base.
NH xxxx—Enter your "Home BBS." (Aids in routing messages to you.)
NQ—Enter your QTH.
NZ xxxx—Enter your ZIP or postal code. (Aids in routing messages to you.)

<1> K4ABT de WB4RHO: at 0917 880921
B,C,D,H,?,I,J,K,L,M,N,R,S,T,U,V,W >

The line above, just to the right of the <1> is the System Prompt. This is the BBS's way of indicating to us that further input is needed so that we may con-

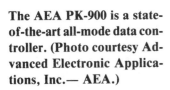

The AEA PK-900 is a state-of-the-art all-mode data controller. (Photo courtesy Advanced Electronic Applications, Inc.— AEA.)

tinue. This line also gives us information about our connection, and it records (saves to disk at the BBS) the time, date, and callsign of our station's connection. The record is maintained in the SYSOP's database, so that he/she may have a record of the BBS activity and the users. (The SYSOP, or system operator, provides and maintains the BBS service for the packeteers.)

I now enter an **LM** to LIST MINE. This will cause a "look-up" table to be activated within the software of the BBS, and a check for mail addressed to my call will take place. If there is mail for me, the BBS will list the "header" of the message, identifying the message number, originator, title, length, time, and date of the message. At this point I can either do an **RM** (READ MINE) or an **R** and the number of the message, and the message will be sent to me. A much easier way is for me simply to send an RM, which tells the BBS software to READ MINE, and the message will automatically be downloaded (sent) to me.

The latest versions of most BBS systems will recognize my call when I connect, and if I have mail in the BBS, the BBS will send me a courtesy message which says "YOU HAVE MAIL WAITING." The same routine that is addressed above is set into motion, and the RM or R #### input from me will bring the message to my terminal. Again, all I need to do is issue an RM and the message will be sent to me automatically. If I'm not sure of the version of the BBS software, then I can issue the LM or RM commands anyway. If there is no mail waiting for me, I will get this response:

(LM)
*** None found.
(RM)
*** None found.
<1> K4ABT de WB4RHO: at 0917 930303
    B,C,D,H,?,I,J,K,L,M,N,R,S,T,U,V,W >

This will get you started using the BBS. Now let's delve deeper into the BBS command structure and feature options.

For help on a specific command, enter **?** **x** where **x** is the command for which you wish help. For example, **? R** will give help for the READ command. Optional fields of the commands are shown inside ( ).

Message commands: (K)ill, (L)ist, (R)ead, (S)end
File commands: (D)ownload, (U)pload, (W)hat
Gateway commands: (C)onnect, (M)onitor

Misc. commands: (H)elp, (?) Same as H, (B)ye, (I)nfo, (J) Who?, (N)ame, (G)et, (T)alk to SYSOP, (V)ersion

H—Give a summary of the commands.
H x—Give an explanation of command x.
H *—Give an explanation of all commands.
B—Log off the mailbox. Disconnecting has the same effect.
Dd file name—Download a file from the mailbox; **d** is the path identifier. Full device and directory path may be given.

Note: The following commands are usually reserved for use by the BBS SYSOP and/or a "remote" SYSOP.

E #—Edit a message header.
EP p—Edit port parameters for port **p**.
ES—Edit system parameters.
EU—Sweep through all users.
EU CALL—Edit a user record.
Fd # FILE opt—Make a file from a message in directory area **d**.
F # FILE opt—Directory path and file name. **Opt:** A—Append to existing file. **Opt: H**—Put the message header into the file.
Jp—Where **p** is a port identifier, give a short list of stations recently heard on that port. The console port list shows the calls of stations recently connected to the mailbox.
K #—Kill message number #.
KM—Kill all messages addressed to you that you have read.
KT #—Kill an NTS message and generate a return "service message."
K@ CALL—Kill all messages @ CALL.
KA CALL—Kill all messages to CALL.
KF—Kill all forwarded but not killed messages.
KF CALL—Kill all forwarded but not killed messages to CALL.
KH—Kill all held messages.

Lists messages in reverse order, newest to oldest. "Personal" messages not to or from you will not be listed.

The various forms of the **LIST** command are:
L—List all new messages since your last log-in.
L #—List all messages back to message #.
LL #—List the last # messages.
L> call—List all messages to this callsign.
L< call—List all messages from this callsign.

L@ call—List all messages addressed at this BBS callsign.

LB—List all bulletins.

LF—List all messages that have been forwarded.

LH—List all held messages.

LK—List all killed messages.

LM—List Mine. Lists all messages *to* you.

LO—List all "old" messages.

LP—List all personal messages.

LT—List all NTS traffic.

LY—List all messages that have been read.

Additional field at end with optional list style. Semicolon means list all information about message.

R #—Read message number **#**.

RH #—Read message number **#**, showing all routing headers.

RM— Read Mine. Read all your unread messages.

Send message type **?** to station **xxxx** at optional BBS **yyy**.

The mailbox will prompt for the message title and then for the message text. End text entry with a **CTRL-Z** or **/EX**.

? is an optional "type" of message, which includes:

B—Bulletin.

P—Personal. Only the addressee can read or list this type.

T—NTS traffic.

The form **SB xxxx [@ yyy] [$BID]** is also available for compatibility with most bulletin-handling systems.

T—Chat with the SYSOP. Any command or return before the request times out will return you to the normal mailbox prompt.

U filename—Upload a file to the name given. For example, UC WESTNET. BBS Reject will occur if file name already exists.

V—Show what version of the mailbox is running.

W—Give a list of directory areas available on the mailbox.

Wd—Give a list of the files in directory area **d**.

Wd ffff.xxx—Give a list of files in directory area **d** that match the given file specification.

As a rule, sending an LL to the BBS will bring the last message entered on the BBS. Sending an LL with a space and a number following it will show you a list of that many messages.

Study the above BBS commands and try your skills with the local BBS. Remember, the BBS is one of our best ways to pass traffic and mail.

NOTES

Chapter 11

Solar Power

Free electricity—that's solar power. We can conserve energy by making use of a utility that is there for the taking. Many times I've heard repeater operators complain because they lost utility power or "there is no power available at the site." If a site is at a good height and there is ample sunlight, then there is available power! There is one minor task to perform, however, and that is the conversion of that energy into a usable form of power. Power comes in different forms and from many sources. The easy part is converting it into the kind of power that will suit your needs.

Although there are many kinds of energy sources—such as water, wind, organic, sunlight, and others—for this discussion I will concentrate on the source that is most available to us on a daily basis.

Conversion of the sun's energy is easy enough, and in the long run solar power is the most economical form of power. If we go about the task of harnessing and storing this energy, we soon discover another benefit that we had not considered before.

Maintenance

In this chapter I'll describe the system that I use here in my QTH, and how I use it in my day-to-day operation, which includes running two HF (all-band, 100 watts each) transceivers, three VHF transceivers, and two UHF transceivers. In addition, the solar system at this QTH is powering five terminal node controllers (TNCs). So what's new? Well, here is the best part of all. The 200 ampere, 14 volt battery bank and the two 2 foot by 4 foot solar panels have supplied these transceivers and TNCs with power for more than a year with no maintenance. Well, maybe I did go out to the solar panel stand and wipe the snow or residue from the solar panels a couple of times.

Many amateurs who visit my QTH ask if I have ever had to supplement the solar panels with a battery charger to have enough power to take me through overcast days. The answer: *absolutely not!* In fact, I have had reserve power in the system to let some of the VHF transceivers run continuously, 24 hours a day, for months, and the system has stayed alive and running well through as much as five days without sunlight.

I know of a system on a mountaintop in north Georgia that has been running for more than two years, and the only maintenance to that ROSE switch has been a periodic visit from the ROSE op to check out the site.

This Siemens M75 is a 47 watt photovoltaic module made with 33 single-crystal silicon square cells, which can produce power in as little as 5 percent of full sun. (Photo courtesy Fowler Solar Electric Inc.)

What About Field Day?

Think about this for a moment: Solar power is not reserved for "fixed" use only. Field-day activity is one of the times when you can put this kind of energy to work and make extra points while doing so.

Building a solar-powered station, packet switch, or even a voice repeater can become a useful and beneficial project. Try it for yourself, for your club, or for the Local Area Network (LAN), and discover how easy it is to put the "free" energy to work for you. You will feel great pride and accomplishment in completing this project.

The Demands That Must Be Considered

As my solar project here at this QTH gathered momentum, I found that the design had some external drivers that I had to consider before I could arrive at the final solution. We will now discuss some of these elements and requirements.

First there is the power supply. The number of items that will be attached to your power supply is the first consideration when buying a power supply at an amateur radio store or hamfest. What size power supply do you need to handle your transceiver, TNC, etc.? There is always the possibility of add-ons later, so include some overhead for that requirement. In this case, we will add an extra 5 amps for the possible add-ons. If the transceiver is a 45 watt output unit and the TNC is an all-mode unit, the following numbers might apply:

Transceiver power requirements—13 volts DC at 9 amps
All-mode TNC power requirements—13 volts DC at 1 amp
Total power for both units—13 volts DC at 10 amps

With the add-on power requirement(s), the current demand can go to 15 amps. If the user adds a 150 watt power amplifier, that power demand suddenly jumps by another 25 amps. Without much arithmetic we can quickly see how the current demand becomes 50 amps.

This same rationale is to be used when designing the solar energy system for a home station or a mountaintop packet switch. When designing this system give yourself some leeway so that you are not pushing the power supply to its limits. That is when you could easily find a need for more than just casual or routine maintenance.

Choosing The Solar Panels

There are absolute maximums that can be derived from solar panels of a given size. Keep the batteries away from the operating position and in a well-ventilated space. Place the batteries off the floor if the floor is concrete or earth.

Use large enough wire with heavy insulation so that current loss in the wire is not a problem. If the batteries are located outside, it is a good idea to place them into some kind of ventilated enclosure to reduce terminal oxidation and for heat dissipation.

Another consideration is the type of battery used. I selected the gel-type lead-acid batteries mainly because this minimizes the degree of battery maintenance required.

Setting The Limits Of The Station Needs

There is a limit that must not—I repeat, *must not*—be exceeded. It is the sum of the power output from the solar panels, the amount of expected daylight, and the amount of daylight that is useful as sunlight to build the power to charge the batteries. For the benefit of the prospective solar switch builder, remember this: You can never outguess the weather—*never*. However, I've seen the system go for five days at a time without sunlight.

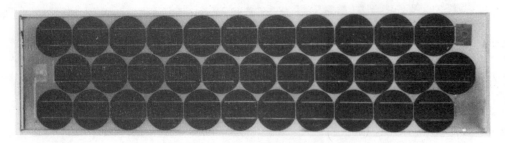

The ARCO 16-2000 is a typical solar module. Solar power does not have to be restricted to fixed station use only. (Photo courtesy Fowler Solar Electric Inc.)

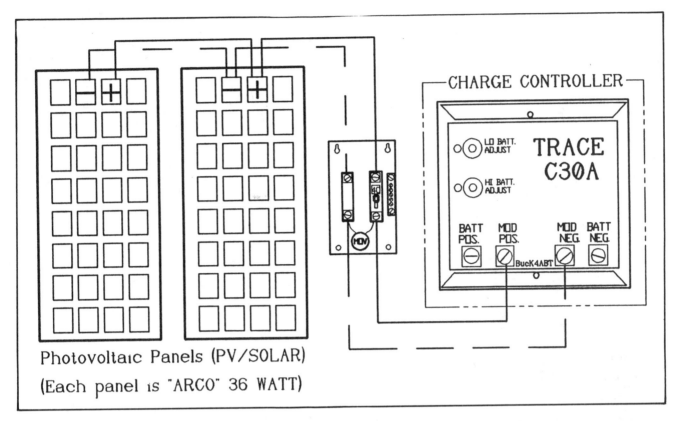

Photovoltaic Panels (PV/SOLAR)
(Each panel is "ARCO" 36 WATT)

Figure 11-1. The charge-controller appears as the second device after the PV (solar) panels. The charge controller and the PV disconnect switch are both grounded to a common ground buss. Notice the metal oxide varistor (MOV) across the disconnect switch posts. The MOV serves two **functions. One is to reduce the amplitude of spikes that may appear on the DC buss, and the other is to trip the disconnect breaker in the event of a "hard" spike or lightning hit.**

Reserve power over extended periods—up to seven days, if possible—as a driver to keep the system operating with little or no sunlight. That number of days might make some folks cringe, but when you have a week without sunlight, you soon learn (the hard way) that it is necessary to have as close to seven days reserve as possible in the power plant.

On the surface, tying the system together may appear to be easy. However, there is a precaution which must be considered and applied before proceeding. When using photo-voltaic (solar panel) devices that are designed to supply charging voltage for a large battery bank, keep in mind the following note of caution: **14 volt solar panels can sometimes generate voltages in excess of 17 volts direct current.** This alone could cause damage to expensive electrical devices.

The "Charge Controller"

Looking at the system flow diagram shown in figure 11-1, we find a device labeled C30A. This device ap-

pears as the second item after the solar (photo-voltaic [PV]) panels and follows the first disconnect panel. It is called a "charge-controller." This device can be the C30A charge controller by TRACE or the M8 by SUN. Here I will describe the M8, and in figure 11-1 I will describe the C30A.

If the M8 is used, connect the brown wire to the battery negative (−) post and the orange wire to the positive (+) post. The manufacturer recommends using a 15 amp in-line fuse in the orange wire. The red wire is attached to the positive terminal of the solar panel, and the white wire is attached to the solar panel negative (−) lead.

If the sun selector M16 is used, it is recommended that a 25 amp fuse be used in the orange wire. The M16 is to be used when the charging currents are above 8 or 10 amps. The maximum charging current should not exceed 20% more than the rated device current-handling capability. There are four LED status indicators on the SUN charge controller. The following lists the definition and purpose/status of each.

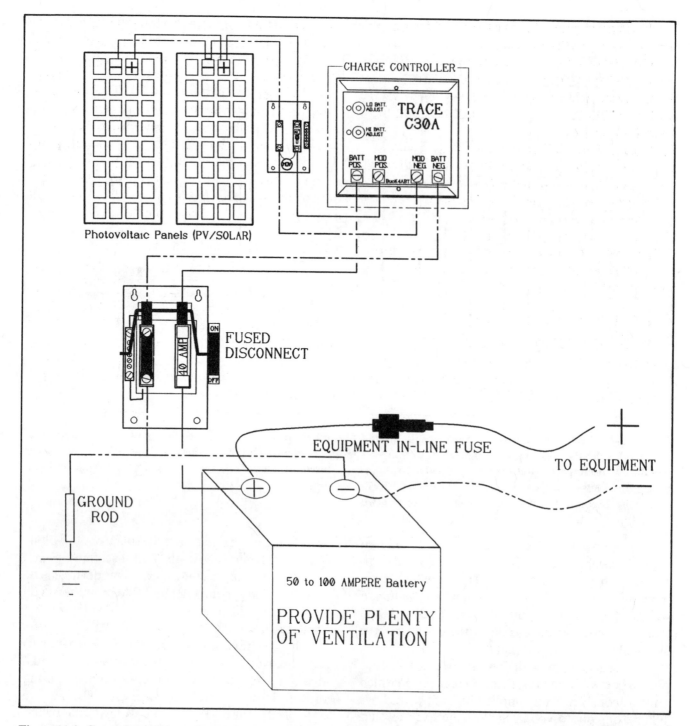

Figure 11-2. Ground the PV panels and all disconnect negative points to a common ground system. The batteries are of the deep-cycle, lead-acid, maintenance-free type. Notice the in-line equipment fuse. When more than one equipment line is needed, it may be necessary to use a DC distribution panel with separate circuit breakers for each unit of equipment addition.

PV READY—Illuminates when the solar panel is emitting sufficient energy to charge the battery.

ANALYZING—Illuminates when the controller has temporarily suspended the charging current to the battery. This is to allow proper chemical (action) mixing inside the battery, which in turn prevents cell damage. In 30 to 60 seconds the charging LED will re-engage.

CHARGING—Illuminates when full charging current is flowing to the battery.

FINISHING—Begins a slow flash rate as the battery reaches full charge. As the battery voltage rises, the flash rate of the LED will increase. This can be used as an indicator to determine battery charge/voltage swing condition.

We learned that the best precaution is to never connect a solar panel directly to the battery without a "governor" to maintain a prescribed level of voltage and current. Now that we have covered the battery considerations, it is time to look at the "load" application.

Before the battery we inserted the "charge controller," which in reality protects the down-stream load (equipment).

"Load Disconnect" Equals Added Protection

In the system we've assembled here, I added a second disconnect panel. Included in this panel are the fuses that are the watch-dog for over-current and in-rush protection for the system (see figure 11-2). After the battery bank I've added another fuse and disconnect panel as further insurance to the system and to allow maintenance protection. Let's call the second disconnect the "load disconnect," as it will remove the load from the batteries while they are being charged to full capacity. When the load is reconnected to the system, it will be at optimum performance. The load disconnect can also be used to provide protection for the battery (or batteries) and other energy-conserving devices up-stream by preventing deep discharge that could cause permanent damage to the batteries.

Dotting The "i's" and Crossing The "t's"

So without great fanfare we have built a digital store-and-forward station on a remote mountaintop where no manmade electrical power existed.

One final precaution: When working with lead-acid or any other storage cells, it is a good idea to allow plenty of ventilation to vent harmful gases during periods of high charge rates.

NOTES

Chapter 12

Where Are The Bauds?

Modems of 9600 baud are being implemented in TNCs around the world. Thus far their implementation has been demonstrated with great success. The most popular modem at present is the G3RUH 9600 baud modem, followed by the K9NG 9600 baud modem.

The James Miller, G3RUH, design is marketed by MFJ, PacComm, and Kantronics. It is easily installed in the TNC of its associated manufacturer: the MFJ-9600 is used with the MFJ products, the Pac-Comm NB96 is used with the PacComm TNCs, and the Kantronics 9600 baud modem is implemented inside the Kantronics DataEngine (DE-56). The MFJ-9600 modem is compatible with the internal disconnect headers in the MFJ TNCs and other TNC-2 clones.

Kantronics has put together a 9600 baud RF modem and transceiver combination. Aside from this combo from Kantronics, there's not much to try for except to join the hackers who are making do with the available choices that can be found in the commercial transceivers. To make this effort work, we must modify the varactor DIRECT FM modulated stages for data input, and connect the receive AF output at the quadrature detector outputs.

Some transceiver manufacturers are planning to introduce in the near future radios that are configured and wired with the tie-points that I mentioned above, brought out to an accessory connector. This kind of support is what we packeteers are looking for. These connections are for the varactor data (FSK) input, and the quadrature detector outputs (before the diodes, please). This is something that will allow the SYSOPs among us to make the needed connections without tearing the guts out of a good transceiver.

Several SYSOPs with whom I'm acquainted are using the MFJ version of the G3RUH modems in the MFJ-1270B TNCs, and attaching them to radios with wide receive and transmit bandpass characteristics. It is preferred that a transceiver to be used with 9600 baud have a 16.5 kHz bandpass, since this is the amount of spectrum that is occupied by the Direct Frequency Shift Keyed (DFSK) transceiver.

The MFJ-9600 modem design is licensed from James Miller, G3RUH, and is in use worldwide. This

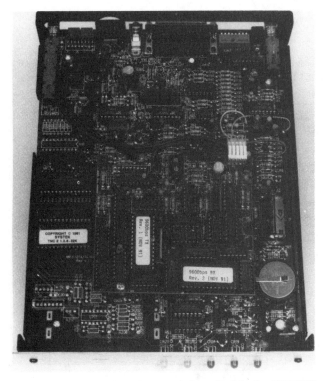

MFJ packet controller shown equipped with the G3RUH 9600 baud modem. (Photo courtesy MFJ Enterprises, Inc.)

Complete details for interfacing and use of the DRSI DPK-9600 baud TNC can be found in the "Packet User's Notebook" column in the May 1993 issue of *CQ* magazine. (Photo courtesy Digital Radio Systems, Inc.—DRSI.)

modem design gives the packeteer a means of creating a flexible transmit wave-form filter design that can compensate for audio differences in many production transceivers. One important feature that stands out in the G3RUH modem and is incorporated in these TNCs is the digital generation of the transmit audio wave-form. The precise shaping limits of the signal bandwidth can be made to tidy up the amplitude and phase response in the receiver dedicated circuits. The result is a compatible filter system within the data detection circuits for optimum data recovery and minimum errors.

We Have Arrived

We can go one of two ways with packet radio. Either we diversify our bands by adding more spectrum (and we know that will not happen), or we get going now and increase the amount of data that can be transferred in a given time period.

Let's Increase The Baud Rates

With the files that are being exchanged by packeteers across the country and the spectrum load that has occurred, the need for a common-sense approach to data handling has arrived. We are passing everything from 256 color, high-resolution pictures and video, to digitized voice that is sampled up to 22,000 times per second. Digital audio requires a large amount of storage space, but even more demanding is the requirement for transmission speed.

Soon we will have the capability of passing both digital audio and digital video at the same time. It does not take a prophet to predict that we will need the increased speed capability so we can take part in the transmission and reception of these powerful mediums. There is a limit to the number of calls and QSOs that can take place on one frequency. Twelve-hundred

baud and 2400 bps have long since been outgrown. The same frequency that is now carrying 5 QSOs (10 target stations) operating at 1200 bauds can carry 8 or 10 times that many QSOs at 9600 baud.

Have I Convinced You Yet?

Alright, here's more. There was a time when Bob Slomka, WD4MNT, and I used MULTICOM.EXE to pass binary files at 1200 baud on 145.61 MHz in central Georgia. Many times these files were over a half-million bytes in length. The time to pass that size file was counted in hours. With 9600 baud we can now pass the same half megabyte of data in less than 15 minutes.

If you plan to implement a 9600 baud modem in your TNC-to-transceiver combination, determine if the transceiver you hope to use employs TRUE or DIRECT FM, and *not* PHASE modulation techniques. The PHASE modulated transceivers are more contrary in a high-speed data transmission application.

Here is a partial list of transceivers that have been modified to operate at 9600 baud:

Alinco: DR-1200 DATA RADIO, DR-110, DR-112; ALR-22, ALR-72, ALR 709.
ICOM: IC series 28A, 38A, 228, 271, 290H, 471, and 3200.
Kantronics: DVR 2-2 DATA RADIO (9600 bps ready when used with the DE-56 DataEngine). The D4-10 can be used with the Data Engine, or combined with other controllers as defined elsewhere in this book.
Kenwood: TR series 751A, 7500, 7700; TM series 211, 212, 221, 231, 431; TS series 700 and 770.
Standard: C58, C140.
Yaesu: FT series 212, 221, 230.

The Alinco DR-1200 Data Radio has an out-of-the-box 2400 baud rate, but can easily be modified to operate at 9600 baud. For details see the "Packet User's Notebook" column in the May 1993 issue of *CQ* magazine. (Photo courtesy Alinco Electronics, Inc.)

The Kantronics Data Engine enables baudrate capabilities up through 56,000 baud. (Photo courtesy Kantronics Inc.)

Although I have made 9600 baud modifications to a few of the transceivers mentioned above, some of these modifications were provided to me by helpful readers of my "Packet User's Notebook" column in *CQ* magazine. Many of these mods were *not* tested by me. Therefore, neither I nor the publisher of this book assume any responsibility for errors or damage resulting from the use of interface and modification information contained in this book. Persons attempting these changes should also be familiar with microcircuit soldering techniques.

Important Note: In some transceivers we've found IF passband limiting caused by the 455 kHz ceramic IF filter. An RC network may be needed to broaden or by-pass this filter so the 9600 bps data can reach the discriminator. An alternative would be to exchange the filter(s) for one (MURATA) with a wider bandpass characteristic.

When installing a 9600 bps modem in a TNC or when connecting the TNC to the transceiver, use *shielded* wire to the transmitter modulator. Use a separate shielded audio wire from the discriminator output for the receive audio.

With the 9600 bps systems we have not been able to interface the transmit and receive audio via the mic input and external speaker output. This is because the audio signals at the mic input and the speaker output have gone through both pre-emphasis and de-emphasis, and the audio may not be data worthy. The average 2 meter transceiver was designed for use in voice applications. To use the same audio connecting points would load the data path with phase shift and distortion, and that just won't work!

As seen in the list of usable transceivers in the preceding section, we have the vehicle to implement these data rates now. So what do we have to do to get a 9600 bps system into operation?

We can start by obtaining the MFJ-9600 or the Pac-Comm NB96. Either of these 9600 baud modems will fit inside the MFJ-1270B, MFJ-1274, or PacComm TNC with the correct disconnect header.

There is another 9600 baud modem called the K9NG modem, but the receiver passband requirements at present are very close. To use this modem, a broad receive passband becomes even more important.

Making The Mods

Remember the statement I made earlier: **Use shielded audio wire for the cables to and from the 9600 baud modem/TNC into the transceiver—one for the transmit and another for the receive, and keep them as short as possible.** Don't try to make it work without the shielded cable, because it won't!

Because many of you will be using the readily available transceivers for 2 meters and above, try to obtain one with TRUE FM. These are usually the transceivers that are crystal controlled, synthesized, and use a varactor modulator. Some are off the shelf, ready to purchase, such as the Alinco DR-1200 and the Radio Shack HTX-202, etc.

We will be connecting our transmit audio from the 9600 baud TNC/modem to the FM-modulated stage of the transceiver. When making this transmit audio connection, it is necessary to use a decoupling capacitor and resistor in series with the audio line.

In the illustrations I've drawn some of the circuits that will help you locate the correct interface points. I've used an RC network in the interface line of both the receive and transmit audio connections. The components in the RC network are identified in the drawings simply as C1, C2, R1, and R2. For clarification, the capacitor (C1) value is in the range of 4 to 8 uF and of the nonpolarized type. Resistor (R1) value can range from 8200 ohms to 15K ohms. The role of the resistor is very important, as it serves to prevent extreme detuning of the varactor modulated stage.

If "padding" of the transmit audio is needed (unlikely), add C2 and R2 from the transmit audio line to ground. (See figure 12-1 A, B, C, D, E, F, G, H.)

Some Kenwood VHF and UHF transceivers are also being successfully used at 9600 bps. TR-751 receive audio connections are made to pin 9 of the TA7761P, or IC2 on the receive unit PC board. Use the 4 uF capacitor and the shielded audio cable we discussed earlier.

Transmit audio is injected to the FM modulator at the junction of resistor R81, varactor diode D21, and

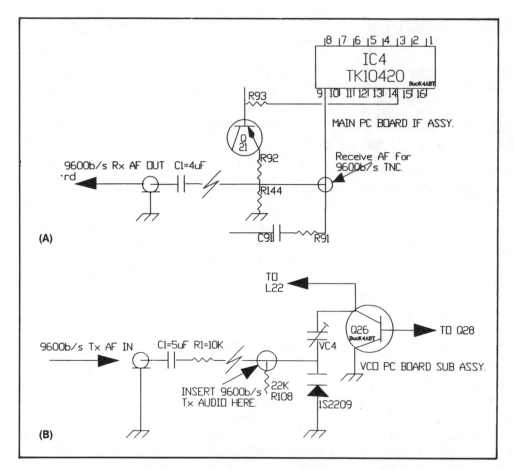

Figure 12-1 (A & B). (A) is similar to the Alinco ALR-72 (UHF) and the ALR-22T (VHF) receive sections. The section of the IF shown is located on the main PC board. (B) is the FM (modulator) section located on the VCO section PC board.

crystal X1 (10.695 MHz). Modifications for both transmit and receive are made on the receive PC board.

Although default parameters in the TNC will work okay, there are several which *must* be changed. Set the **TXD**elay between 10 and 14, or to no more than 150 milliseconds maximum. The **FRACK** can be set at 4 to 7 seconds, depending on channel congestion, 7 being heavy traffic and 4 being clear, or little use.

Time To Plug-'n-Play

Since much of what is now occurring in digital communications is happening at the higher speeds, there is a trend towards looking at easier ways and means of implementing our backbones. The only problem is that of compatibility. The Kantronics TNC does not (yet) support TheNet or the ROSE code. Yet the Kantronics D4-10 does present us with a transceiver that has the high-speed analog data port ready to "plug-'n-play." In addition, I've had the opportunity to use this transceiver in a moderately high RF environment with pleasing results. With the Kantronics D4-10 in mind as a ready-to-operate 9600 baud radio,

then all that is left is the addition of a TNC that is ROSE or TheNet (EPROM) compatible and equipped with a 9600 baud modem.

Both ROSE and TheNet are two of the more popular protocols. However, we will use the ROSE hardware configurations for this project. This may also catch the eye of some other transceiver manufacturers and convince them take a closer look at what is needed in the packet world. I'm hoping they will soon come up with easy access to the tie points inside the transceivers. The tie points we hope they will give us are "to" the varactor stage(s) of the true FM transceiver (transmit audio) and "from" the quadrature detector and/or discriminator (receive audio). The push-to-talk line is another point that will need to be accessed via these external connectors.

Let's Get Busy

The input and output lines to and from each of these circuits must be shielded from each other. There is no common shield for both signals. I repeat: Each signal *must* have a separate shield (notice the transmit and receive audio lines inside the TNC in figure 12-2).

Figure 12-1 (C & D). (C) is similar to the PC board of the Alinco DR-110, DR-112, and the DataRadio 1200. (D) shows the solder side of the VCO PC board. Transmit audio is injected at the junction of R11 and varactor diode D2.

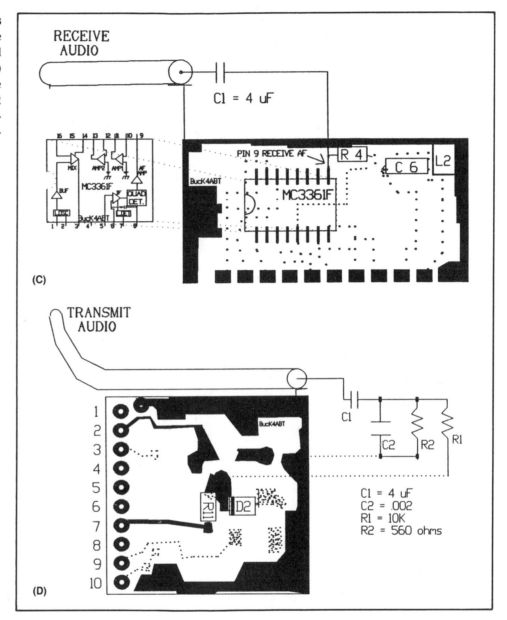

If these lines are not properly shielded (separately), the data within these lines can be corrupted and destroyed by noise ingress from surrounding circuitry. Don't forget to widen the IF filters, as a wider bandpass is needed with 9600 baud and higher.

Phil Anderson, WØXI, and the design team at Kantronics in their wisdom presented us with the following choice(s), or better said, a set of building blocks to formulate the system we might need. On their D4-10 ten watt UHF transceiver they gave us easy access to the varactor modulator and the receiver discriminator output with both TTL or analog connectors *on the rear panel.*

The TNC will be the MFJ-1270B equipped with the MFJ-9600 baud modem and the ROSE code.

Radios Capable Of Higher Data Speeds

When constructing the interface cable that connects the TNC to the D4-10 (figure 12-3), it is necessary that mini-coax or shielded cable be used on both the transmit and receive audio lines. This means a separate line for each, *not* both lines in a common shield. Tie the shields to pin 2 of the TNC exit connector only. *Do not* ground these two cable shields at the transceiver

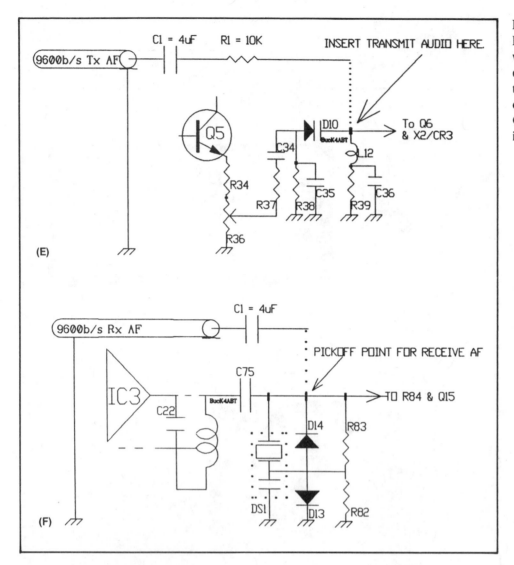

Figure 12-1 (E & F). (E) is the IC-25A modulator section which shows the transmit audio injection points. (F) identifies the location where receive audio may be extracted. C1 is user supplied. The value is between 4 and 8 mFd.

end of these lines. In addition, it is wise to use ferrite beads over the center conductor of each line at the transceiver end of each line.

I was able to use the connector and cable that were included with the MFJ-1270B TNC for the interface cable between the Kantronics D4-10 transceiver and the TNC. The shield in this cable is soft copper distributed around the signal-bearing lines inside the cable.

Hardware Configuration

If your needs are for a network backbone or trunk, then you may wish to build the transition from a LAN frequency to a UHF backbone. If this is the case, a transceiver/TNC combination is necessary for the LAN frequency, and another transceiver/TNC combination is needed for the high-speed backbone side of this gateway.

It will be necessary to construct the gateway interface cable illustrated in figure 12-4 and set the async port dip switches on both TNCs to complement each other. The end result of this gateway configuration should allow connects similar to the one shown in figure 12-5. This is a simplified approach, but the results

The Kantronics D4-10 is a UHF data-ready radio capable of up to 56Kb. (Photo courtesy Kantronics Inc.)

Figure 12-1 (G & H). (G) indicates an easy to locate point to extract 9600 baud data. Notice the pick-off point is located at the junction of R61 and R62. (H) indicates the 9600 baud audio injection point.

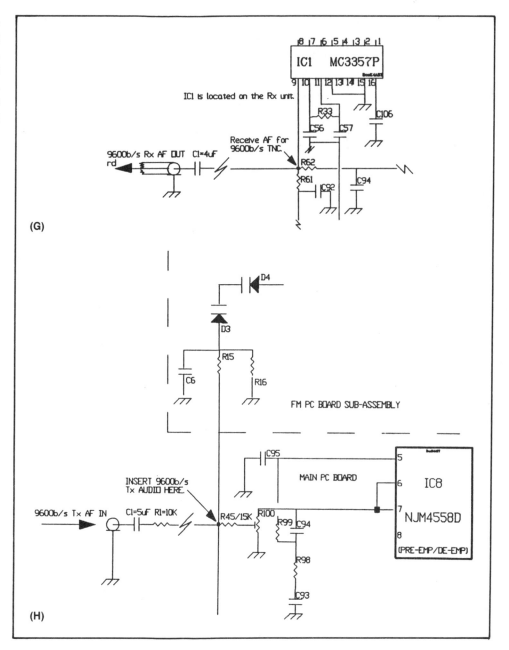

can be the removal of heavy traffic from our LANs and the movement of this traffic to the high-speed trunks. Make sure that your TNC is equipped with the modification shown in figure 12-6. After studying the application of our gateway, you will notice that the local user does not experience the heavy traffic that is usually seen across the system.

This implementation of high-speed trunks can be a joint venture by other systems, such as BBS forwarding. If the LAN is set aside as an emergency or keyboard-to-keyboard-only system, there must be a means of barring access to the BBS via the LAN and backbone. A means of limiting access to BBSes from

a keyboard-to-keyboard-only system is supported within the ROSE firmware. Likewise, the same feature is supported from the BBS side of the high-speed backbone.

Through cooperation among SYSOPs, the backbone can be utilized to move vast amounts of traffic over long paths without affecting the throughput at the user level. Again, I caution the SYSOP who might consider mixing various kinds of traffic. *Please* consider the end result before leaping into this line of fire. Experience has taught some of us that mixing CONFERENCE, CONVERSE, BBS forwarding, and DX spotting nodes and networks into one system will bring

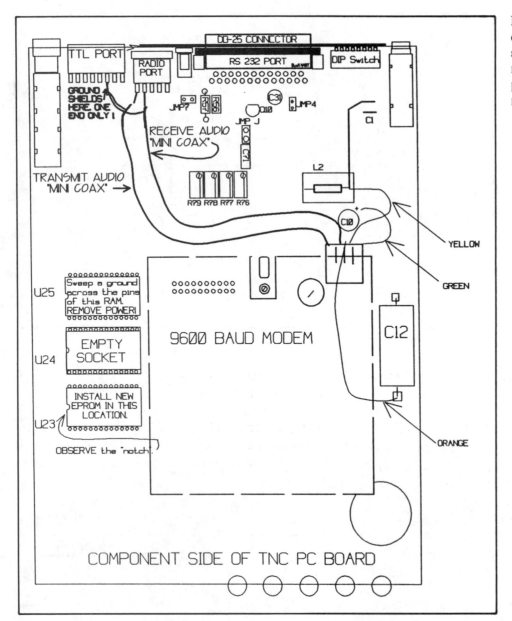

Figure 12-2. TAPR TNC-2 clone showing the installation and wiring of the 9600 baud modem. This TNC may be purchased with the 9600 baud modem already installed.

the system to its knees. It is a matter of experience that has taught us to use separate frequencies for these multi-connect CONFERENCE and spotting clusters.

When an application calls for integration of several baudrates and frequencies, the necessity for more than two TNCs emanates. I've included the means to augment our multi-port gateway and baudrate transition in figure 12-7.

Speed Calls For More Speed

One final tip for the high-speed switch SYSOP: When burning the EPROM for the TNC, I found that us-

ing a faster (150 nanosecond) 27C256 improved the throughput over the 200 and 250 ns EPROMs that we were able to get away with in the 1200 baud switches.

In the current ROSE code there are several PAR files that support RF or USER levels ranging from 1200 to 9600 baud and via the async and radio ports. Be sure to look at the PAR files and determine which one you wish to use. In my case I wrote my own to fit my application. By experimenting you can tailor the 9600 baud parameter files to give optimum throughput. If you don't care to experiment, you may wish to use the K4ABT4.PAR file shown in table 12-1. Table 12-2 is the K4ABT4.BAT file I use to execute

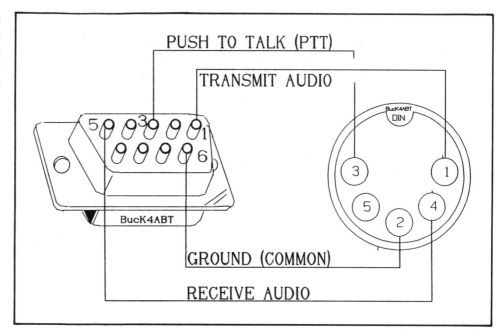

Figure 12-3. The Kantronics D4-10 UHF DataRadio interfaced to the TNC-2 or clone equipped with the G3RUH (MFJ-9600 or PacComm NB-96) baud modem.

the K4ABT4.PAR file. All examples are included in the ROSE 3.0 ZIP file.

Module and Component Vendors

For more information or to order from Kantronics, you may contact your Kantronics dealer, or Kantronics directly at 1202 E. 23rd Street, Lawrence, Kansas 66046. Their telephone order line is 913-842-7745. BBS @ 300, 1200, 2400, N,8,1 is 913-842-4678.

The TNC-2 type clones MFJ-1270B (TNC) and MFJ-9600 (9600 baud modem) are available from MFJ Enterprises, Inc., Box 494, Mississippi State, Mississippi 39762. Their order line is 800-647-1800.

A similar TNC-2, 9600 baud TNC is available from PacComm Packet Systems, Inc., 4413 N. Hesperides Street, Tampa, Florida 33614. Telephone 800-223-3511.

A TNC-2 compatible, full-blown 9600 baud TNC, the DPK-9600, and the DPK-2, a 1200 baud TNC-2 compatible, are available from Digital Radio Systems, Inc. (DRSI), 2065 Range Road, Clearwater, Florida 34625. Telephone 813-461-0204.

The Ultimate "Time Machine"

One half of the time machine is the Kantronics D4-10, 430 MHz transceiver. It supports both wide- and narrow-band switch-selectable filters to support both high-speed data and voice communications. The second half of the time machine is the Kantronics Data Engine.

What is the speed of the data? Well, that option is the user's choice. Coupled with the Kantronics dual-port Data Engine, with a CPU built around the 80196 (V40), we can travel from 1200 baud to 19,200 baud and beyond, depending on the modem that is installed onto one of two port headers inside the Data Engine.

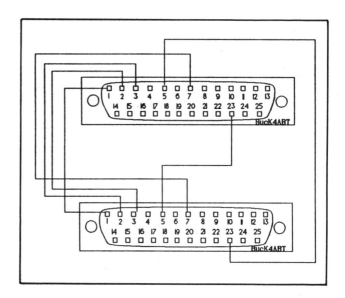

Figure 12-4. This two-port (async) connector is used to bridge (gateway) two ROSE switches that are on different frequencies or different baudrates. (Note: Pins 2 to 3 and pins 5 to 23 are rolled [inverted] between the two connectors.)

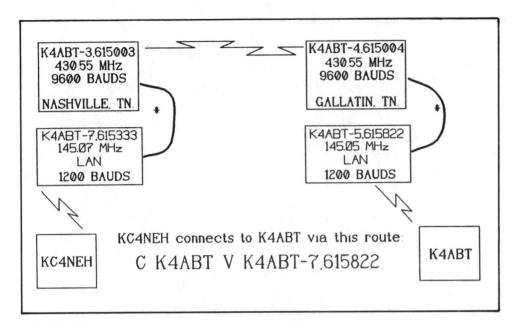

Figure 12-5. The * indicates the cable shown in figure 12-3. Set the dip switches on both TNCs for 9600 baud via the async ports. Some older TNCs may not handle 9600 baud. In any case, set the async port speed on both TNCs to the same speed and the highest usable baudrate.

The Data Engine can support two modems and can address each port separately, or it can be outfitted with modems that allow an access port for local user speeds, while the other port is addressing the high-speed backbone link to rapidly move data across a large network. The SYSOP (and in many applications, the user) can configure the Data Engine to his or her liking.

There are other systems which can handle similar high data rates, but they are built around a number of component blocks that heterodyne, mix, and double to reach the frequency and baudrate needed to handle the kind of traffic load required here. In addition, the cost is almost prohibitive when you consider putting one of the other systems on a mountaintop for a link, because we have to consider the added cost of a PC or compatible to be used as a controller.

With the Kantronics D4-10 and the Data Engine equipped with the 19,200 baud modem we have a compact, simple to install and maintain link that consumes

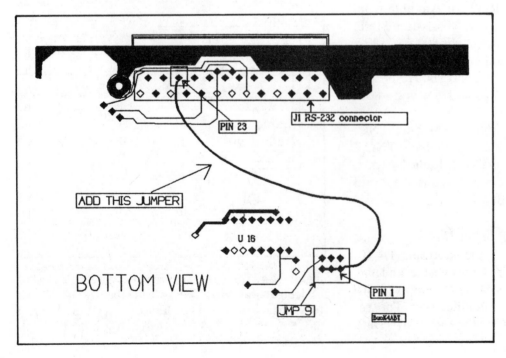

Figure 12-6. As with other TheNet and ROSE gateway switches, this minor modification is necessary when used in a back-to-back gateway into a backbone application. (Note: This modification is already installed in the MFJ-1270B and MFJ-1274 1992 production units.)

Figure 12-7. This diode matrix is used to interface up to four ROSE switches in a gateway configuration. Switches may be used to link to other frequencies, baudrates, or trunks. This matrix may be used to interface a LAN to a high-speed backbone.

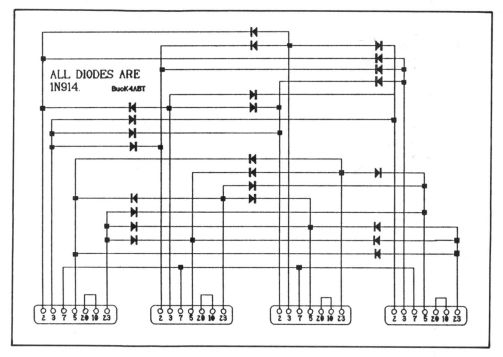

```
OUT = K4ABT4.ROM
CALL = K4ABT-4
DIGI = K4ABT4
ADDRESS = 3100404040
SETUP = STANDBY while your call is setup
COMP = Your call is complete to
BANner = 430.55 MHz; 9600 bps ROSE network, limited
    access trunk.
#For user ports and trunks via RF above 9600 baud.
#Pound sign (#) is for comment lines only. Not an instruction.
L3FRack 1
L3RESPtime 1
L3MAXframe 5
L2CHeck 500
L3CHeck 1800
L2FRack 1
L2RESPtime 4
L2MAXFrame 5
L2RETry 12
L3RETry 12
0TXDelay 70
```

Table 12-1. By experimenting you can modify the 9600 baud parameter files to give you optimum throughput. This table illustrates a new means incorporated into the "MAKEPROM.EXE" program to configure the EPROM using a "PARameter" file that is driven by the BATch file shown in table 12-2. The text above was saved as K4ABT4 .PAR. If you don't care to experiment, you can use this file.

very low current and occupies less than one square foot of space. This is a networking tool that takes the headache out of putting the packet hobby into the 21st century.

There is no question about it. We are entering a new era of packet communications. We SYSOPs have long hoped for an economically priced package that would make our upgrade to the higher speeds less of a burden. When we speak of economy, we have to consider the large amount of data that can be handled on a backbone when utilizing a speed of 19,200 bauds per second. By that statement, I mean we can move massive data loads across a metropolitan area for BBS forwarding (and for several BBSes) with ease.

The addition of user traffic input to the backbone

MAKEPROM/INCLUDE = K4ABT4.PAR
This short line of text is saved as K4ABT4.BAT.
This BATch file executes the PARameter file at table 12-1 into the MAKEPROM.EXE program to produce an EPROM when loaded and run in the EPROM burner.

Table 12-2. This is the K4ABT4.BAT file used to execute the K4ABT4.PAR file shown in table 12-1.

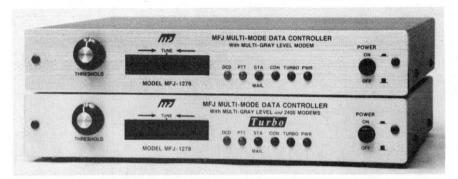

On top in the photograph is the MFJ-1278 multi-mode data controller, a TNC capable of many modes of data communication. The Turbo model shown on the bottom is capable of 300, 1200, 2400, or 9600 baud operation. (Photo courtesy MFJ Enterprises, Inc.)

via another (or second) Data Engine port/gateway accessed at 1200 baud is easily combined with the BBS traffic stream. The data rate is so fast on the backbone that a user on the LAN side of the system should not notice any slow down at all in the throughput. Even with our AX.25 packet mode using half duplex, to see a data file exceeding 10,000 bytes in length passing from one station to another in less than 10 seconds is a thrill to behold. We may have arrived at a way to support our backbones and trunks for a long time to come.

Are We Having Fun Yet?

Putting it together was almost too easy, and putting the two stations into operation was just as easy. At no time in my packet career have I ever seen anything go together so smoothly. In one evening I had both ends of the 19,200 baud links up and running. It's easy to answer the question "Are we having fun yet?" Yes, we are having fun!

The D4-10 is a product that carries its own weight when it comes to design features. It has performed well in a room where an RF rich VHF and UHF environment exists. The D4-10 handles a TXDelay of seven with room to spare, and it appears to be fairly linear with respect to the amplitude and phase. The D4-10 supports both analog and TTL data input. For the purpose of our application we will be using the TTL input port of the D4-10 transceiver.

The D4-10 is a two-channel, crystal-controlled transceiver, and it is shipped with 430.55 MHz crystals installed in the channel one position. The power output of both of our D4-10's was measured. Although the D4-10 is rated at 10 watts output, both our units exhibited more than 12 watts output into the AAE 440 MHz quads. The purpose of the 6-element 440 MHz quads was to traverse the 25 mile path from Gallatin to Nashville, Tennessee without adding an amplifier. It worked!

With so much already written to cover the design theory of these systems, I'll not bore the reader with more coverage of the same material. Why invent the wheel again when it has already been invented? Let's use the already invented wheel in a real-world application. This is not a review of the product. It is an explanation of how we make use of a new tool that is already designed and ready for use in packet radio.

Preliminary Requirements

Because the modem for this application is designed to operate at either 9600 or 19,200 baud, it may be used with other transceivers which will only handle up to 9600 baud. The manufacturer therefore ships the modem with the strapping options set to the 9600 baud staking pins. Our plan was to use the Kantronics Data Engine Packet controller and the D4-10 transceiver in the wideband mode, and at 19,200 baud. For openers, remember to set the three pushbuttons at the right side on the front of the transceiver to the relaxed, or OUT, position when using 19,200 baud. Only the ON/OFF pushbutton will be engaged, or ON (in).

If you are adding or installing the DE-19K2/9K6 modem, follow the instructions in the manual for easy installation of the modem. Make a note of the port on which you installed it, or better yet, keep notes as you go.

When you have the modem in place, the next step is to move the four strapping options that apply to 19,200 baud to their respective staking pins. See figure 12-8 for the location and position of the "strapping options." The options in figure 12-8 as shown are set for the 19,200 baud position.

Begin by setting **JP9** for **center pin to pin "B."** "A" to center pin is the 9600 baud setting, and center pin to pin "B" is the 19,200 baud setting.

JP12 determines the mode of the receive data. It has a jumper from the **center pin to the "B" stake.**

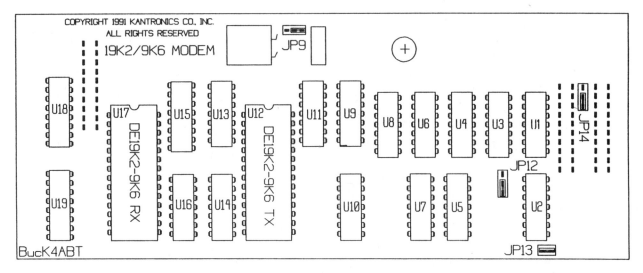

Figure 12-8. The Kantronics modem as shipped is set for 9600 baud. Set jumpers JP9, JP12, JP13, and JP14 to the positions shown here. (See text for detailed setup and configuration.)

This is the TTL position. When the center pin (stake) is jumped to the "A" pin, the modem is set for analog receive data input.

JP13 is a two-stake jumper, and *it is strapped* when using 19,200 baud. According to the modem manual, this option is the DCD setting for 19,200 baud when the jumper is in place. However, when looking at the schematic, it appears to be part of an RC filter network which in effect places two 100K resistors in parallel when the jumper is in place. It may be that when in place, the jumper increases the DCD response time when operating at 19,200 baud.

JP14 located near the modem headers is the counterpart to JP12. This option determines whether the transmit data is analog or digital (TTL). Set this option (jumper) between **center pin and the "B" pin.**

Building The Interconnect Cables

Before I go too deep into the interfacing of the D4-10 and the Data Engine, I need to pass on a hint about the actual soldering to the wires inside the multi-conductor cables. After the outer covering was removed and the bundle of wires inside exposed, I discovered I did not need to strip away the insulation to make the ends of the wires bare. I touched the ends of the wires with the hot soldering iron, and the insulation shrank away from the end of the wire, exposing enough of the bare wire to apply solder and attach it to the connector pin(s). Be careful not to overheat the wires, or you may expose more wire than desired or needed.

In figure 12-9 I've drawn the DB-15 Data Engine connector to the DB-9 connector of the D4-10 transceiver input. The view is of the rear, or solder, side of the connectors as they appear when connected to the radio and Data Engine, looking at the rear of each unit.

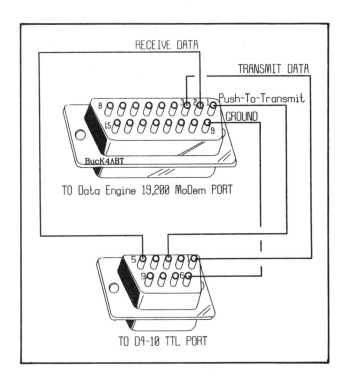

Figure 12-9. The Kantronics Data Engine to D4-10 interface cable.

Be sure to use the six-wire shielded cable that is supplied with the Data Engine. This is necessary to preserve the integrity of Part 15, Sub-Part "J" of the FCC certification and to prevent ingress of external electro-static noise into the perishable data inside the lines.

For our 19,200 baud application only four of the six wires are used. The signals and the pin numbers for each connector are shown in table 12-3.

In figure 12-10 I've drawn the interface cable for the Data Engine interconnection to the computer. The drawing applies to a normal configuration of the PC or compatible serial ports which employ the DB-25 or DB-9 connector(s). Table 12-4 supports the same connections, but the colors may vary from those shown in table 12-4.

The 80486 machines I use are capable of 38,400 baud. However, the terminal program I used would only handle up to 19,200 baud terminal speed. Set

Data Engine (DB15 Port)		Transceiver "D4-10" (DB9 Port)	
Transmit Data Pin 3	<TO>	Transmit Data Pin 1	
Receive Data Pin 2	<TO>	Receive Data Pin 5	
Push-To-Talk Pin 1	<TO>	Push-To-Talk Pin 3	
Signal Ground Pin 9	<TO>	Signal Ground Pin 6	

Table 12-3. Interfacing the DE (Data Engine) to the D4-10, 430 MHz transceiver.

your terminal program to 19,200 baud if you plan to transfer files from computer to computer via this system.

The Final Configuration

To run tests of the 19,200 system prior to relocating one of the systems to the top of a tall building in downtown Nashville, I operated them into 15 watt dummy loads. Here is where I ran into the first and

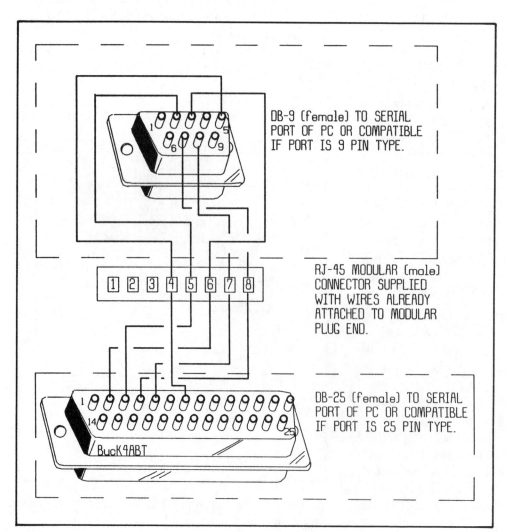

DB-9 (female) TO SERIAL PORT OF PC OR COMPATIBLE IF PORT IS 9 PIN TYPE.

RJ-45 MODULAR (male) CONNECTOR SUPPLIED WITH WIRES ALREADY ATTACHED TO MODULAR PLUG END.

DB-25 (female) TO SERIAL PORT OF PC OR COMPATIBLE IF PORT IS 25 PIN TYPE.

Figure 12-10. This drawing shows both the DB9 and DB25 connectors interfaced to the RJ-45 cable and connector supplied with the Kantronics Data Engine. (See text and table 12-2 for detailed interface information.)

Data Engine, Data Port			Computer Serial Port	
(RJ-45)			(DB25)	(DB9)
1 = DSR	*(BLUE)	<TO>	NC	NC
2 = DCD	*(ORANGE)	<TO>	NC	NC
3 = DTR	*(BLACK)	<TO>	NC	NC
4 = SGnd	*(RED)	<TO>	7	5
5 = RcvData	*(GREEN)	<TO>	3	2
6 = TxData	*(YELLOW)	<TO>	2	3
7 = CTS	*(BROWN)	<TO>	5	8
8 = RTS	*(GRAY)	<TO>	4	7

Table 12-4. Interfacing the Kantronics Data Engine (data port) to the PC serial COMPort.

only obstacle of this project. The RF output of the D4-10's is outfitted with BNC female connectors. All connectors from the dummy loads and my antennas are fitted with the standard UHF or PL-259 connectors. The project halted for an hour while I drove into Hendersonville to Radio Shack to purchase two SO-239 to BNC male adapters. Radio Shack calls them ''RF Adapters.'' The Radio Shack Catalog part number is 278-120.

Having told you how to solve that problem, I can now tell you how to finish setting up the system to make a connect. Attach the dummy load(s) and prepare for action!

Connect the Data Engine(s) to the computer(s) and turn them ON. Be sure to set your terminal and comport parameters to communicate with the Data Engine. Once you've found a way to strike the * (asterisk) quickly enough, you will receive the prompt ENTER YOUR CALL. Input your callsign, and you are ready to start the final configuration.

The parameters of the 19,200 baud system can be set to the user's liking. However, I would suggest using the following settings to start. As you fine-tune the system for your backbone, trunk, or network, you may discover a configuration that will better favor your application.

At the CMD: prompt set the following parameters and commands:

TXDelay to 6 to 9
DWait 0 (zero)
PERSIST 200
SLOT 1-2
RESP 2-3

Set different callsigns into each system, and then make a connect and begin having fun ''in the fast lane''—packet-wise, that is. Amaze yourself by transferring a few 100,000 byte files between the 19,200 baud stations. Don't turn away for more than a minute or two, or you will miss a 100 kilobyte file transfer.

If you are listening on the frequency, here is the next thing you are going wonder about: Is it really transmitting packet? There is a living, breathing animal in that box. If it's hissing, it's there!

This system can be configured in several ways, so I've listed the various components of this system separately so you may order the configuration that suits your application or station requirements. For more information or to order this system contact Kantronics directly.

Summary

The topics we've covered in this manual represent only the beginning of all the enjoyment you can have in the hobby of packet radio. But for a moment, let's stop and consider the following statement: **Your packet station could prove to be your most worthwhile asset!**

Packet radio can serve many needs and purposes. However, one application for the use of packet stands out from the rest. Your packet station can provide one of the most effective and efficient means of communications in situations regarding the safeguarding of life, health, and property. The recent hurricanes and earthquake disasters were but a few of the times when packet radio excelled and proved that it is indeed one of the most effective traffic-handling mechanisms available to the modern amateur.

With this book we hope we have given you the tools to build your packet station, get it up and running, and begin having fun ''digitally''!

NOTES

Defining RS-232

In the computer world there exists a communications standard. This standard is called RS-232. We can describe RS-232 in this manner: interfacing Data Terminal Equipment (DTE) to Data Communications Equipment (DCE) employing serial, binary data exchange.

A connector that is often associated with the RS-232 standard is referred to as the DB25 connector. Late-model computers use the DB9 connector.

Let's familiarize ourselves with the DB25 connector (figure A-1). At figure A-2 we have the DB9 connector configuration. It is easy enough to remember

the name of both connectors, because each label defines the number of pins associated with the connector. Of these early 25 signals/pins, there were only 10 of them needed when interconnecting the TNC to a terminal. Today only 5 of these signals are used. In applications where ''software handshaking'' is implemented, we use as few as three of these signals.

Here are the ten pins:

1—Equipment Ground
2—Transmit Data
3—Receive Data
4—Request To Send
5—Clear To Send
6—Data Set Ready
7—Signal Ground
8—Data Carrier Detect
20—Data Terminal Ready
22—Ring Indicator

The five most often used signals (hardware handshake):

2—Transmit Data
3—Receive Data
4—Request To Send
5—Clear To Send
7—Signal Ground

For three-pin (software handshake) applications:

2—Transmit Data
3—Receive Data
7—Signal Ground

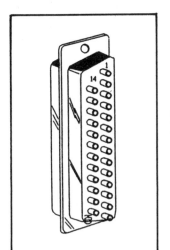

Figure A-1. Shown here is the DB25 connector sometimes associated with the RS-232 communications standard for computer serial ports.

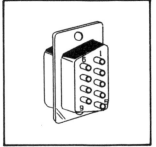

Figure A-2. The DB9 (DE9) connector found on many late-model PC compatibles. This is another connector that is often associated with the RS-232 signal format.

Note: I use pin 1 only when both the TNC and terminal are attached to the same power outlet. In a few instances there have been cases where power outlets were served by different power mains, with different ground locations. A potential difference caused the small (pin 1) wire in the cable (equipment ground) to become heated, and worse. Using pin 1 to preserve the integrity of the equipment ground is okay, but it is wise to test the DTE and DCE for possible differences in potential before attaching these lines.

Another Approach To "Ten Pins"

Let's take a pin-by-pin tour of the ten pins with which we are working.

Pin 1 is the equipment ground and may be used to ground one end of the shielded cable, if used. It is often advisable to use shielded cable in the presence of RF. As many of us are aware, the environment around an amateur station will support this reasoning.

Pin 2 is the transmit data (TD). This is defined as the data line from the DTE to the DCE. A good rule of thumb to remember is that all signals are identified from the DTE or terminal end.

RS-232 standard sends the transmitted data to the DCE on pin 2, while the received data is fed to the DTE from the DCE on pin 3.

Pin 3 is Receive Data (RD) line from the DCE to the terminal (DTE).

Pin 4 is the Request To Send (RTS), and is the complement to pin 5, Clear To Send (CTS). The RTS is issued from the terminal to the TNC, but no data will be sent until a CTS is issued by the DCE (TNC).

Pin 5 is the complement to pin 4, since pin 5 is responsible for the CTS signal from the TNC to the terminal. These two pins are the keys to the most often used technique of hardware handshaking. When under hardware control, the terminal will not send until a CTS is received from the DCE.

Pin 6 is the "data set ready" and is complemented with pin 20 and pin 8. Pin 6 indicates the TNC ready, or powered, and is used in some packet BBS applications.

Pin 7 is connected to the audio signal shield and in general is used as the data, timing, and control reference grounds.

Pin 8 is the signal detector and is related to pins 6 and 20. Sometimes called Data Carrier Detect (DCD). This type of signaling is generally related to the telephone modems.

Pin 20 is the final complement to pins 6 and 8. It is also used with telephone modem installations. It confirms a condition of phone off hook to pin 6. In many applications pin 20 and pin 8 are tied together to assert that DTE is on.

Pin 22 is sometimes used as the ring indicator, but more often the control "G" will fill the ticket where packet radio is in use.

From these 10 pins we find that only pins 2, 3, 4, 5, and 7 are really necessary for our day-to-day packet use.

CAUTION: Some DCE/TNCs may use pin 25 to carry + 12 volts to the TNC from an external source. When the DCE/TNC-provided power source is used, the internal + 12 volts will appear on this pin. As a precaution, I never attach a wire to pin 25.

A Message For ROSE SYSOPs

H ere is a short document that I've written to enable designated remote SYSOPs to assist the Network Manager in maintaining the latest configuration and feature files in the ROSE switches. This is necessary, since many of them are on the outlying areas of the network. There are other files which are included on a disk that is used by the remote SYSOPS.

ROSE Switch Uploading And Configuration Procedures

If you are connecting to a local ROSE switch to upload the configuration tables, follows these steps:

1. Connect to the switch LOADER—example:

C LOADER V [SwitchCall],[SwitchAddress]
<enter>
(C LOADER V K4ABT-6,912987) <enter>

The display will appear similar to the following:

CMD:***CONNECTED to LOADER VIA
    K4ABT-6,912987
    Call being Setup
    Call Complete to LOADER-0 @
    3100912987
    ROSE X.25 ROSE switch Version
    90#### by Thomas A. Moulton W2VY

2. After you receive "CALL COMPLETE to LOADER-0," etc., send a colon and ten zeros (:0000000000) to LOADER:

:0000000000 <enter>

If this is the first time you have configured the ROSE Switch, you should receive the following:

Entry #0 LOADER - Application Boot Interface
    Version 1.1
OK

3. Always check the LOADER to see that the feature you are uploading is not already present. *Never* send the same feature file to a ROSE switch LOADER if it is already present! As an example, if you send the colon and ten zeros (:0000000000), the following display is returned to you from the ROSE switch:

Entry #0 LOADER - Application Boot Interface
    Version 1.1
Entry #1 CONFIG - ROSE X.25 Packet Switch
    Configuration Interface Ver 2.2
OK

It will *not* be necessary to upload the CONFIG.LOD file to the LOADER.

4. If CONFIG is *not* present in the LOADER, then CONFIG.LOD must be uploaded to the LOADER. After the CONFIG.LOD file upload is complete, there should be *three* OK's on the screen. This indicates that the CONFIG.LOD upload was successful. DISCONNECT from LOADER of this switch!

5. To continue with the configuration of the new ROSE switch you should connect to the newly loaded CONFIG feature of the switch. In this example I will use the same ROSE switch as before, except this time we will be connecting to CONFIG as follows:

C CONFIG V K4ABT-6,912987 <enter>

The display reads:

CMD:*** CONNECTED to CONFIG VIA
    K4ABT-6,912987
Call being Setup
Call Complete to CONFIG-0 # 3100912987
ROSE X.25 switch Version 90#### by Thomas A.
    Moulton W2VY
OK
OK
OK
OK
OK
OK
OK
OK
OK
OK
OK

6. At this point you may begin loading the FILE-NAME.TBL file for the *ROSE switch to which you are connected*. This table is generated with the F9 key in the TERMINAL mode, by C.EXE. The FILE-NAME.CNF file contains the configuration you wrote earlier using the editor in BUXTERM.EXE. (See SYSOP's Manual "ROSESYS.TXT.") In my case, I assign a name to the .CNF file which will identify the ROSE switch to which it is to be sent.

Note: Be *absolutely positive* that you are sending the correct file to the corresponding ROSE switch. Observe the title line at the top of the ALT R window in BUXTERM.EXE to confirm file name!

In the example above, I receive the CALL COM-PLETE TO CONFIG-0, etc., and I begin the upload to the ROSE. The title of the table for this ROSE switch is "K4ABT6.TBL."

7. When the K4ABT6.TBL upload is complete, I wait until the complete file is sent. To let me know the upload is a success, the ROSE switch will send me eleven "OK's."

8. DISCONNECT now! Your configuration upload is complete. To upload the feature files, connect to the LOADER and upload the INFO.LOD, MHEARD.LOD, and USER.LOD files. *Never upload* the ZAPIT.LOD file unless there is reason to RESET the ROSE switch or to break a "locked" ROSE switch.

The BOOTER.LOD will clear all feature files from the switch. *It will not erase* the current routing tables. To erase the routing tables, simply upload a new routing table to the switch.

Glossary

Because the terms and jargon used when "talking" packet have not yet developed into a group of buzz words, the new user may find it best to learn some of the speech descriptors that are used in place of single-word identifiers.

Every prospective packet user needs a place to start. There are two basic requirements the beginner needs to know. The first is an understanding of the jargon, or buzz words, and terms associated with packet radio. The second requirement is directly related to how a system is put together.

Through many years of experience as the Manager of Engineering for a large network of television and radio broadcast stations, I learned the following very important lesson. Those prospective technicians and engineers who had first had a few months of on-the-job training earned a higher grade average in their subsequent school training than those who had not first been exposed to the practical work environment. To this end and in that same manner I wish to address this glossary. I will first give a synopsis of the packet station requirements and configuration procedures. Later we will study the single words or letter groups which are called "acronyms and mnemonics."

Here is what I hope we will achieve. When you read this section, you will begin to learn about packet in a new way and with more understanding. The rest of the learning process is easy because the nature of packet is such that it allows learning by osmosis.

After completing this glossary, you will have a better understanding of the dialog of packet radio and of the terms used throughout this manual. You will discover that you have developed a more fluent understanding of packet radio and the terms associated with your hobby. In addition, you will be able to put your packet station on the air with less difficulty than those who have not studied this kind of material. You will become the expert of packet jargon in a very short time, and will have an increased knowledge of the subject. You will be on the air with the best of us who enjoy the most advanced communications mode to date.

Terms and Definitions

ABaud Data speed between the computer terminal and the TNC.

ACK Acknowledgement from target packet station receiving an error-free packet from the station that sent the original packet of information.

AFSK Audio Frequency-Shift Keying. When associated with VHF packet, the shift is usually 1000 Hz.

ALIAS The "alias" is normally used to identify the location of a digipeater. Many "digis" are given the local airport identifier or the abbreviation for the city or state of origin—i.e., the alias of the Dothan, Alabama digi is "DHN." This also eliminates the need to type in long callsigns and SSID.

ASCII American Standard Code for Information Interchange. The ASCII 7-bit code represents 128 characters, including 32 control characters. Also known as the ANSI standard X3.4 1977. The ASCII code set is used in almost all computers and peripherals. ASCII is the basis of most information which is transmitted by amateur packet stations. The ASCII 7-bit code represents 128 characters, including 32 control characters.

baud The unit of digital signal speed, equal to the number of events per second. Baud is not necessarily the number of bits per second.

BBS Bulletin Board System.

binary A two-state numbering system represented by the 0 (zero) and 1 (one). A binary digit is called a "bit." Digital data is represented by a one or zero bit when used in packet communications.

bit Notation for a binary digit. The smallest unit of digital information. The bit can represent a choice of a one or zero (mark or space) in digital communications.

boot Boot, reboot, and system boot refer to a "cold start" of the computer or related devices.

buffer A storage area or device (normally in software RAM) where data overflow is contained until RAM or disk space can be made available for storage. The buffer is mainly used to hold data while it is being transferred from one computer to another.

CCITT Consultative Committee of the International Telephone and Telegraph. An international counterpart to the Electronic Industries Association in Washington, D.C. EIA standard RS-232 and CCITT V.24 are very similar.

CMD: An on-screen prompt which is displayed by the TNC. This prompt informs the user that the TNC is in the COMMAND mode.

connected The condition that occurs when two packet stations are described as being "connected." The state in which the two stations read only those packets from each other when MCON is OFF.

CTRL C or **Control C** This is used to bring a TNC to the CMD: mode from the CONVERSE or TRANSPARENT mode. It is executed by holding the CTRL (control) key down and pressing the C key. In most TNCs it is not necessary to press <enter> after executing a CTRL C.

CPU The Central Processing Unit which controls data flow and "thinking" function within a computer. When supported by a math coprocessor, it can perform computations at greater speeds.

DB-9P and **DB-9S** Electronics Industries Association's recommended connector for use with RS-422A standard.

DB25P and **DB25S** The connectors that will support all 25 of the RS-232 signals. Recommended by the EIA.

DCE Data Communications Equipment is the device or TNC which provides signal conversion so that data communications can be established, maintained, and discontinued. Some DCE are controlled through local or remote software commands.

decibel (dB) The logarithmic unit or measure of a ratio between two powers, P1 and P2. The equation is:

$$dB = 10 \log_{10} P2/P1$$

digipeater A store-and-forward "digital repeater" which will receive and transmit a data packet on the same frequency. All amateur packet stations are capable of digital repeating in a simplex environment.

dip Dual in-line package, as applied to sub-miniature switches and monolithic integrated circuits.

DTE Data Terminal Equipment is another name for a device or computer that sends and receives data in digital form at its input/output port.

flow control The process that starts and stops terminal output to prevent loss of characters or data by the receiving device.

FSK Frequency Shift Keying. When associated with HF (300 baud) packet, the shift is 200 Hz.

gateway When associated with packet radio, a gateway is a "bridge" that provides a means to digipeat from one frequency into another or from one baud rate to another.

HBaud Data speed between the TNC and the transceiver. Sometimes referred to as the "station to station" baud rate.

HDLC High-level data link control. The process used in X.25 and AX.25 to format data into packets. These packets of data have the destination address, checksum count, and other necessary components added through HDLC to help make it an error-free mode.

host The mainframe computer or massive memory storage facility where accessible data-bases are held. These data-bases are accessed by computers or terminals which are allowed access via preassigned passwords or callsigns.

JHeard A command associated with the PBBS and mailbox features of many popular TNCs. When the JHeard command is invoked, a list of the most recent, or "just heard," stations will be displayed. If the JHeard command is executed with a "J<space>L" the paths indicating the stations callsign, origin, and digipeaters will be displayed.

LAN Local Area Network. When associated with amateur radio packet, the term defines an area or locale where a group of packeteers use the same frequency to converse or receive messages into their personal packet mailbox.

NOVRAM Non-Volatile Random Access Memory

is a memory chip which contains its own power source and holds the present memory, even if the power is removed from surrounding circuitry. External commands provide a means to change the memory.

PBBS Personal Bulletin Board System.

PMS Personal Message System (see also PBBS).

protocol The rules for maintaining communications between similar devices, as with AX.25 maintaining orderly error-free data flow and data link control.

QPSK Quadrature Phase-Shift Keying.

RAM Random Access Memory. That part of a computer or TNC that is holding data, or memory, during the power ON period. If the RAM has "battery backup," the memory remains in the RAM until it is changed. If the RAM has no battery backup, no memory remains when power is removed.

RS-232 A set of signals accepted as a standard by the Electronics Industries Association (EIA) and designed to make the interfacing of computers and networks easier.

RS-422A, RS-423A, and **RS-449** Recent standards which were developed to overcome the defects of RS-232.

SSID The SSID is the specific number applied to the callsign of a digipeater or second, third, etc., packet station. An example is the digipeaters of WB4RHO, and the associated SSIDs located in central and south Alabama, which are listed below.

WB4RHO (2400 bps BBS MBL bulletin board system)

WB4RHO-1 (145.010 MHz) Kennedy mountain
WB4RHO-2 (223.400 MHz) Abbeville, Alabama
WB4RHO-3 (145.010 MHz) Headland, Alabama
WB4RHO-9 (145.090 MHz) Abbeville, Alabama

TAPR Tucson Amateur Packet Radio, a non-profit research group dedicated to the advancement of amateur digital communications.

TCP/IP Transmission Control Protocol/Internet Protocol. The KISS mode activates this mode in some TNCs. This mode is not supported by software for all computers and terminal node controllers (TNCs).

TNC Terminal Node Controller is the combined modem and assembler/disassembler. The interface device between the computer terminal and RF transceiver. The TNC assembles and disassembles packets and provides error detection.

TTL Transistor-to-Transistor Logic. Some TNCs will not accommodate this form of interface. A separate signal converter is required when a TNC does not support TTL. Some TNCs have built-in TTL strapping options or connectors.

TVRO Television, Receive Only. The acronym is used to denote the home satellite receiving system.

WeFAX Weather facsimile, reconstructed satellite pictures and photographs. The WeFAX receive mode is now an added feature of the all-mode digital controllers.

NOTES

Transceiver-To-TNC Interfaces

T36   ICOM IC-290 to DRSI DPK-2
T37   ICOM IC-228 to DRSI DPK-2
T38   KDK FM-2033 to AEA DSP-1232 and 2232
T39   KDK FM-2033 to DRSI DPK-2
T40   KDK FM-2033 to MFJ-1270B, 1274, and 1278/Turbo
T41   KDK FM-2033 to PacComm TNC-200 and Tiny 2
T42   KDK FM-144 to MFJ-1270B, 1274, and 1278/Turbo
T43   KDK FM-144 to AEA DSP-1232 and 2232
T44   KDK FM-144 to DRSI DPK-2
T45   KDK FM-144 to PacComm Tiny 2 and Micro Power
T46   Kenwood 2550 and 2570 to DRSI PC*Packet Adapter
T47   Kenwood TR-2550 to AEA PK-88
T48   Kenwood TS-450S (HF) and TM-231A (VHF) to PacComm PC-320
T49   Kenwood TS-440 and 940 to Dual-Port AEA DSP- 2232
T50   Kenwood TR-7400A to AEA PK-87
T51   Kenwood TR-7600 to Kantronics KAM and KPC-3
T52   Kenwood TS-450S and 950S to AEA DSP-1232 and 2232
T53   Kenwood TR-7600 to AEA PK-88
T54   Kenwood 7950 to AEA PK-232 MBX
T55   Kenwood TS-950S to DRSI PC*Packet Adapter
T56   Kenwood TM-231 to AEA PK-232
T57   Kenwood TS-440/940 to AEA PK-232 (HF)
T58   Kenwood TS-440S/940S ACC-2 to Kantronics KAM HF Port
T59   Kenwood TS-440S/940S to MFJ-1278 (HF)
T60   Kenwood TS-450S/950S to AEA PK-232 (HF)
T61   Kenwood TS-450S ACC-2 to KAM HF Port
T62   Kenwood TS-690S ACC-2 to Kantronics KAM
T63   Kenwood TS-711A ACC-2 to KAM HF Port
T64   Kenwood TR-851A to AEA PK-88
T65   Kenwood TS-950S ACC-2 to Kantronics KAM (HF) Port
T66   Kenwood Handhelds to PacComm Tiny 2
T67   Kenwood TR-7400 to Heath HK-21
T68   Kenwood Handhelds to MFJ-127# Series
T69   Kenwood Handhelds to Kantronics KPC's
T70   Kenwood Handhelds to PacComm HandiPacket
T71   Kenwood TH-25 to HandiPacket
T72   Kenwood TR-50 to AEA PK-88
T73   Yaesu FT-101Z/ZD to Kantronics KAM (HF) Port
T74   Yaesu FT-212RH to AEA PK-88
T75   Yaesu FT-227R to AEA PK-88
T76   Yaesu FT-2700 and 4700 to DRSI DPK-2
T77   Yaesu FT-2700R to AEA PK-88
T78   Yaesu FT-290R to DRSI DPK-2
T79   Yaesu FT-290R to AEA PK-88
T80   Yaesu FT-301D to MFJ-1278 (HF)
T81   Yaesu FT-747 to Kantronics KAM (HF)
T82   Yaesu FT-747 and 757 to MFJ-1278/Turbo
T83   Yaesu FT-747 and 757 to AEA DSP-2232
T84   Yaesu FT-757 GTX II to Kantronics KAM (HF) Port
T85   Yaesu FT-7#7 Series to MFJ-127# Series

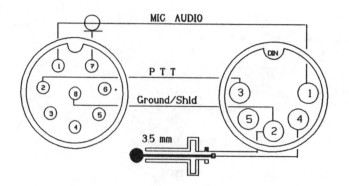

ALINCO DR-110 TO DRSI DPK-2

T1. Alinco DR-110 to DRSI DPK-2.

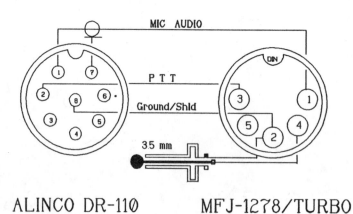

ALINCO DR-110   MFJ-1278/TURBO

T2. Alinco DR-110 to MFJ-1278/Turbo.

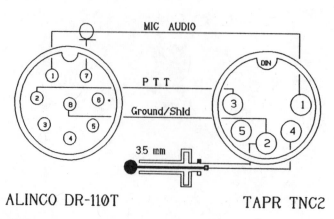

ALINCO DR-110T   TAPR TNC2

T3. Alinco DR-110 to TAPR TNC-2.

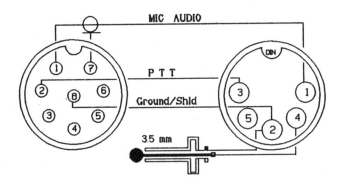

ALINCO DR-110   MFJ-1270 & 1274

T4. Alinco DR-110 to MFJ-1270B and 1274.

(Note: Receive AF may be taken from the external speaker or from pin 6 of the transceiver mic connector.)

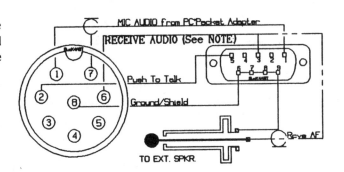

ALINCO DR-1200   DRSI PC·Packet Adapter

T5. Alinco DR-1200 to DRSI PC*Packet Adapter.

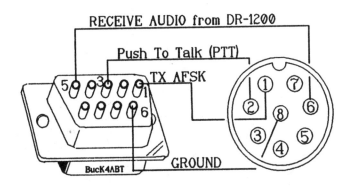

KPC-3   ALINCO DR-1200 VHF Transceiver

T6. Alinco DR-1200 to Kantronics KPC-3.

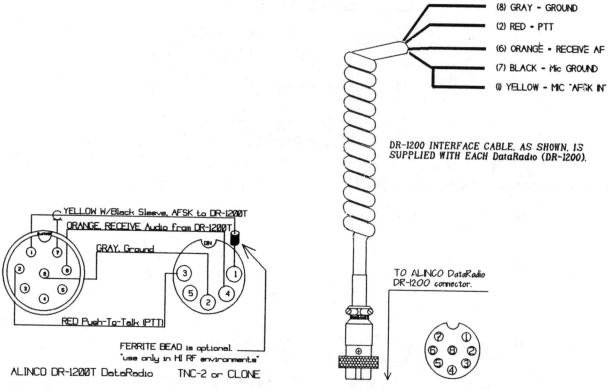

(8) GRAY - GROUND
(2) RED - PTT
(6) ORANGE - RECEIVE AF
(7) BLACK - Mic GROUND
(1) YELLOW - MIC "AFSK IN"

DR-1200 INTERFACE CABLE, AS SHOWN, IS
SUPPLIED WITH EACH DataRadio (DR-1200).

TO ALINCO DataRadio
DR-1200 connector.

YELLOW W/Black Sleeve, AFSK to DR-1200T
ORANGE, RECEIVE Audio from DR-1200T
GRAY, Ground
RED Push-To-Talk (PTT)
FERRITE BEAD is optional.
"use only in HI RF environments"

ALINCO DR-1200T DataRadio   TNC-2 or CLONE

T7. Alinco DR-1200 Interface Cable to TNC-2. An 8-pin connector with cable attached is supplied with the Alinco DR-1200T DataRadio. Pins 3, 4, and 5 are not used in the connector. Therefore, only five wires extend from the TNC end of the interface cable.

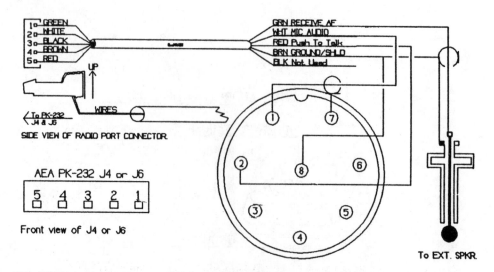

1 GREEN
2 WHITE
3 BLACK
4 BROWN
5 RED

GRN RECEIVE AF
WHT MIC AUDIO
RED Push To Talk
BRN GROUND/SHLD
BLK Not Used

UP
To PK-232
J4 & J6
WIRES
SIDE VIEW OF RADIO PORT CONNECTOR.

AEA PK-232 J4 or J6

| 5 | 4 | 3 | 2 | 1 |

Front view of J4 or J6

To EXT. SPKR.

AEA PK-232 PakRatt   ALINCO DR-570, DR-590 or DR-599 VHF/UHF TxCvrs

T8. Alinco DR-590 and 599 to AEA PK-232/MBX and PakRatt.

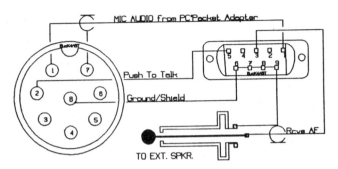

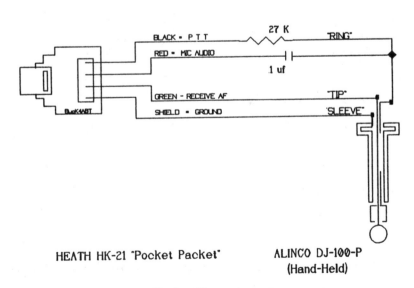

T9. Alinco DR-110 to DRSI PC*Packet Adapter.

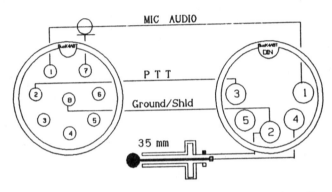

T10. Alinco ALR-22 T/E to MFJ-1278 (either port).

T11. Alinco DJ-100-P to Heath HK-21 ''Pocket-Packet.''

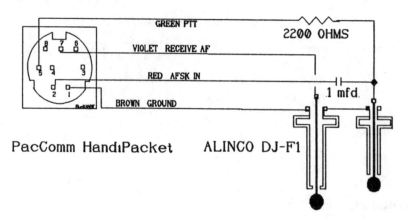

T12. Alinco DJ-F1 to PacComm HandiPacket TNC.

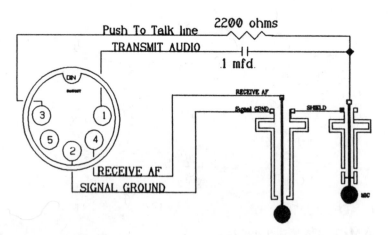

T13. Alinco DJ-F1 Handheld to TNC-2 or clones.

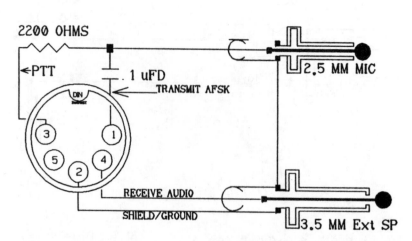

T14. Alinco DJ-580 Twin Band to DRSI DPK-2.

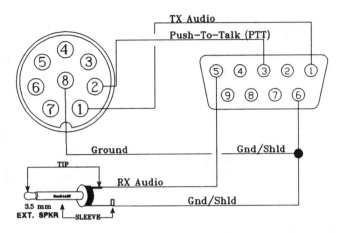

ALINCO DR-110/DR-112   KANTRONICS KAM/KPC

T15. Alinco DR-110/112 to Kantronics KAM/KPC.

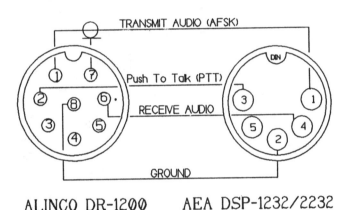

ALINCO DR-1200   AEA DSP-1232/2232

T16. Alinco DR-1200 to AEA DSP-1232/2232.

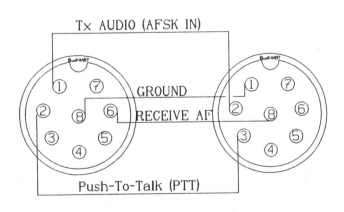

ALINCO DR-1200   AEA PK-88

T17. Alinco DR-1200 to AEA PK-88.

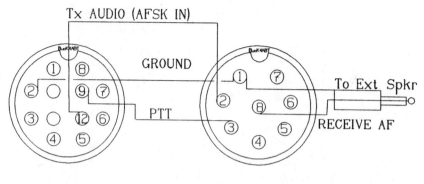

T18. Azden 3000 to AEA PK-88.

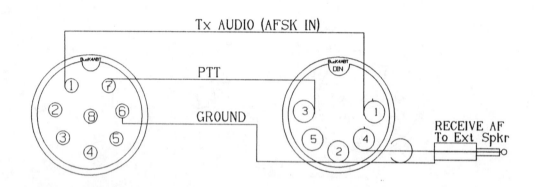

T19. Azden PCS-4000 to MFJ-1270B, 1274, and 1278/Turbo.

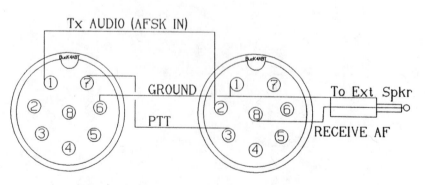

T20. Azden PCS-4000 to AEA PK-88.

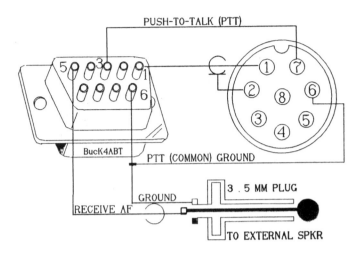

KANTRONICS KAM   AZDEN PCS-4000/5000

T21. Azden PCS-4000 and 5000 to Kantronics KAM.

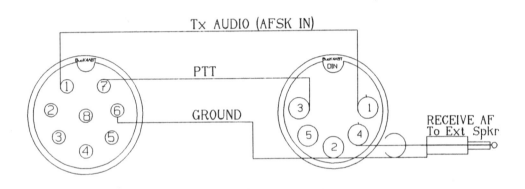

AZDEN PCS-5000   DRSI DPK-2

T22. Azden PCS-5000 to DRSI DPK-2.

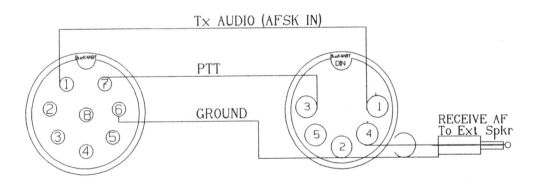

AZDEN PCS-5000   PacComm Tiny-2 & MicroPower

T23. Azden PCS-5000 to PacComm Tiny 2 and Micro.

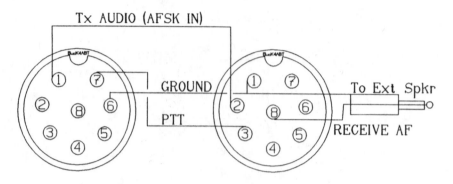

AZDEN 5000   AEA PK-88

T24. Azden PCS-5000 to AEA PK-88.

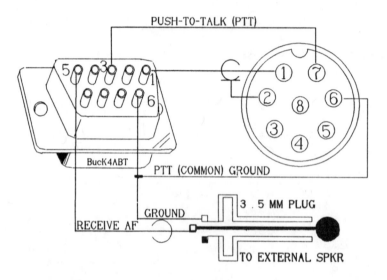

KANTRONICS KAM & KPC-2400   AZDEN PCS-7000(H)

T25. Azden PCS-7000(H) to Kantronics KAM.

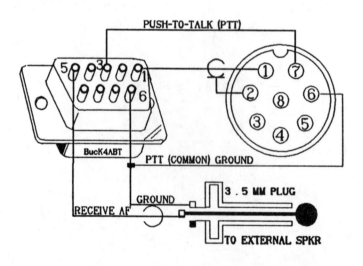

KANTRONICS KPC-3   AZDEN PCS 7000(H)

T26. Azden PCS-7000(H) to KPC-3.

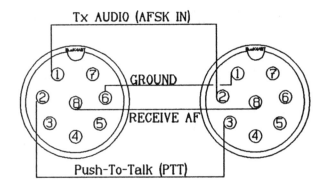

ICOM IC-27A, 28A, & 38A   PK-88

T27. ICOM IC-27A, 28A, and 38A to AEA PK-88.

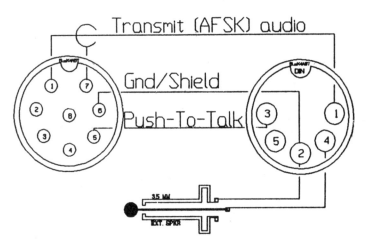

ICOM IC-290   PacComm Tiny 2

T28. ICOM IC-290 to PacComm Tiny 2.

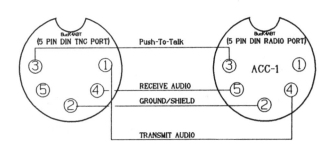

MFJ-1278 WITH PARAMETERS SET FOR HF.   ICOM IC-735.

T29. ICOM IC-735 to MFJ-1278 (HF).

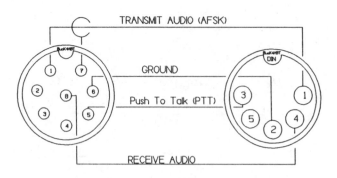

ICOM IC-27, 28, 38, & 735   AEA DSP-1232

T30. ICOM IC-27, 28, 38, and 735 to AEA DSP-1232.

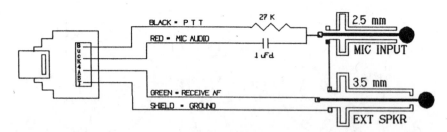

HK21 "Pocket-Packet"   ICOM IC-02AT Handi-Talki.

T31. ICOM IC-02AT to Heath HK-21.

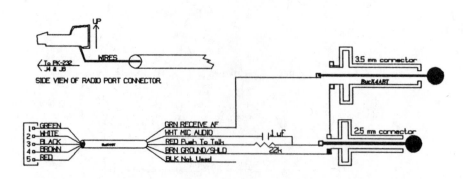

AEA PK-232 Multi-Mode Controller   ICOM HT's.

T32. ICOM handhelds to AEA PK-232 MBX.

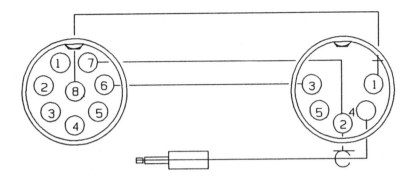

ICOM 27A. & 27H   TNC-2 or CLONE

T33. ICOM IC-27A and 27H to TAPR TNC-2 or clones.

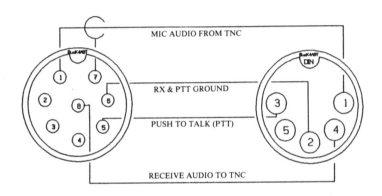

ICOM IC-28A/H   DRSI DPK-2

T34. ICOM IC-28A/H to DRSI DPK-2.

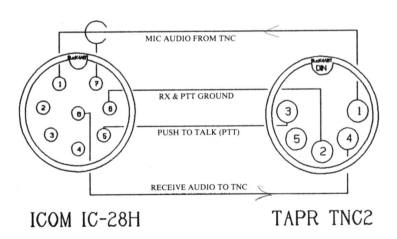

ICOM IC-28H   TAPR TNC2

T35. ICOM IC-28H to TAPR clone.

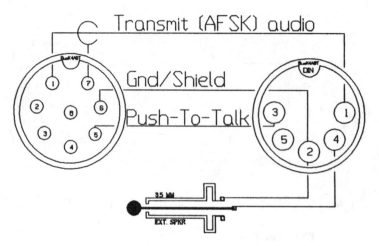

ICOM IC-290   DRSI DPK-2

T36. ICOM IC-290 to DRSI DPK-2.

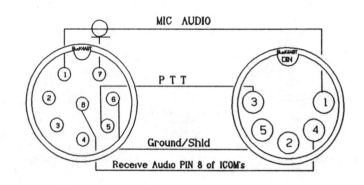

ICOM IC-228   DRSI DPK-2

T37. ICOM IC-228 to DRSI DPK-2.

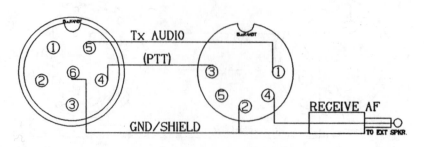

KDK FM-2033   AEA DSP-1232, & 2232

T38. KDK FM-2033 to AEA DSP-1232 and 2232.

TRANSCEIVER-TO-TNC INTERFACES   117

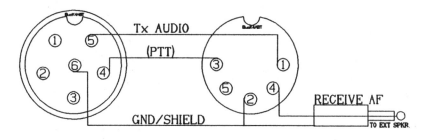

KDK FM-2033   DRSI DPK-2

T39. KDK FM-2033 to DRSI DPK-2.

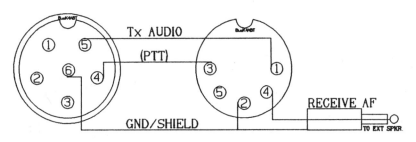

KDK FM-2033   MFJ-1270B, 1274, & 1278

T40. KDK FM-2033 to MFJ-1270B, 1274, and 1278/Turbo.

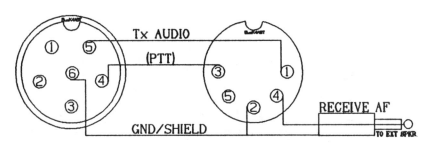

KDK FM-2033   PacComm TNC-200, & TINY 2

T41. KDK FM-2033 to PacComm TNC-200 and Tiny 2.

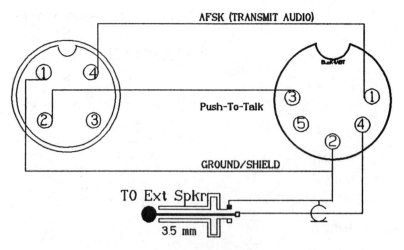

KDK FM-144   MFJ-1270B, 1274, & 1278/Turbo

T42. KDK FM-144 to MFJ-1270B, 1274, and 1278/Turbo.

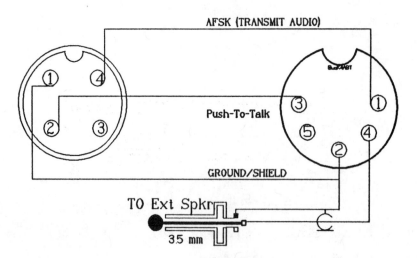

KDK FM-144   AEA DSP-1232 & 2232

T43. KDK FM-144 to AEA DSP-1232 and 2232.

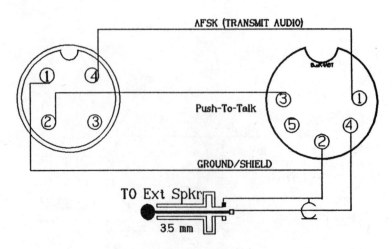

KDK FM-144   DRSI DPK-2

T44. KDK FM-144 to DRSI DPK-2.

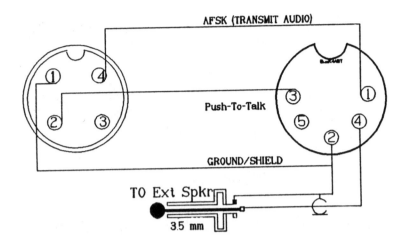

KDK FM-144   PacComm TINY-2 & MicroPower 2

T45. KDK FM-144 to PacComm Tiny 2 and Micro Power.

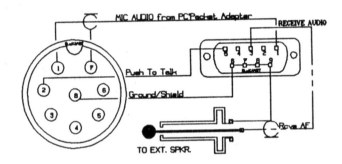

KENWOOD 2550 & 2570   DRSI PC*Packet Adapter

T46. Kenwood 2550 and 2570 to DRSI PC*Packet Adapter.

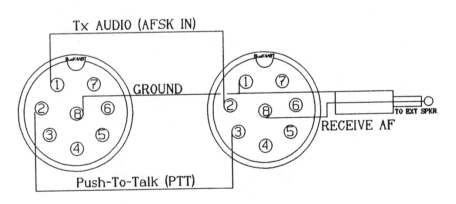

KENWOOD TR-2550   AEA PK-88

T47. Kenwood TR-2550 to AEA PK-88.

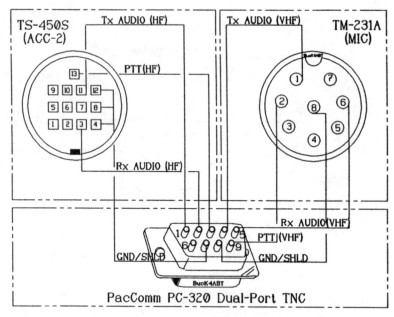

KENWOOD TS-450S (HF) & TM-231A (VHF)   PacComm PC-320

T48. Kenwood TS-450S (HF) and TM-231A (VHF) to PacComm PC-320.

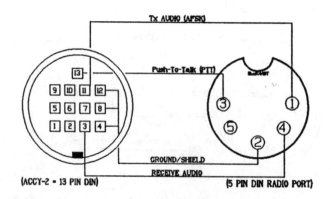

KENWOOD TS-440 & 940   AEA DUAL-PORT/GATEWAY DSP-2232

T49. Kenwood TS-440 and 940 to Dual-Port AEA DSP- 2232.

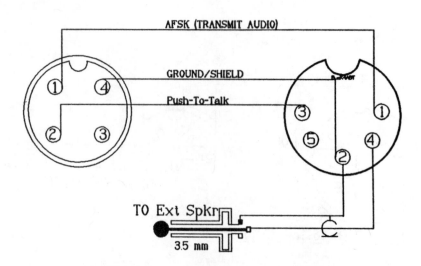

KENWOOD TR-7400A   AEA PK-87

T50. Kenwood TR-7400A to AEA PK-87.

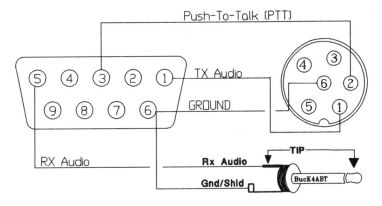

KANTRONICS KAM, KPC-2, & KPC-3   KENWOOD TR 7600

T51. Kenwood TR-7600 to Kantronics KAM, KPC-2, and KPC-3.

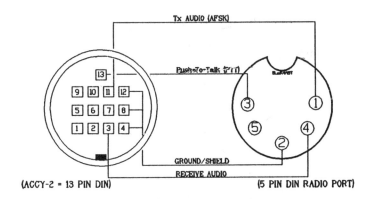

KENWOOD TS-450S & 950S   AEA DSP-1232 & DSP-2232

T52. Kenwood TS-450S and 950S to AEA DSP-1232 and 2232.

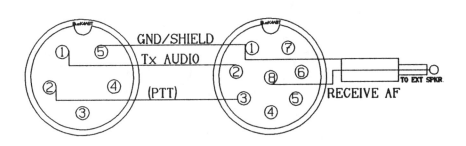

KENWOOD TR-7600   AEA PK-88

T53. Kenwood TR-7600 to AEA PK-88.

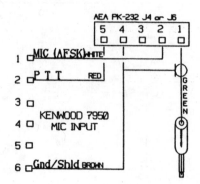

KENWOOD 7950   AEA PK-232 MBX

T54. Kenwood 7950 to AEA PK-232 MBX.

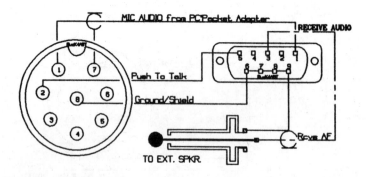

KENWOOD TS-950S   DRSI PC*Packet Adapter

T55. Kenwood TS-950S to DRSI PC*Packet Adapter.

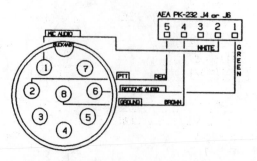

KENWOOD TM-231 transceiver   AEA PK-232 MULTIMODE

(Note: Receive AF is available at pin 6 of the mic connector.)

T56. Kenwood TM-231 to AEA PK-232.

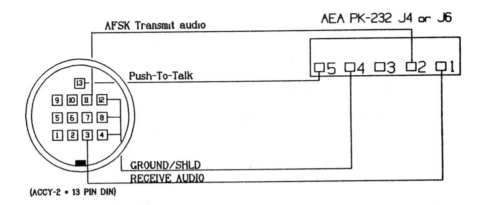

KENWOOD TS-440/940   AEA PK-232; PORT SET FOR HF OPERATION.

T57. Kenwood TS-440/940 to AEA PK-232 (HF); port set for HF operation.

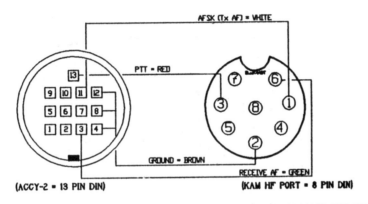

KENWOOD TS-440S/940S ACC-2   KANTRONICS KAM HF PORT

T58. Kenwood TS-440S/940S ACC-2 to Kantronics KAM HF port.

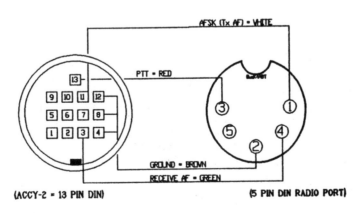

KENWOOD TS-440S/940S   MFJ-1278 PORT 1 OR 2

T59. Kenwood TS-440S/940S to MFJ-1278 port 1 or 2 configured with parameters set for HF operation.

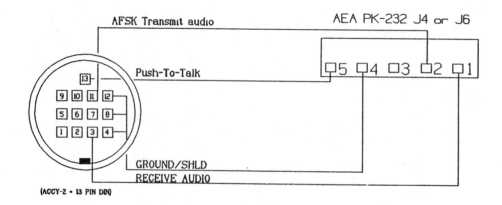

KENWOOD TS-450S/950S   AEA PK-232, PORT CONFIGURED FOR "HF"

T60. Kenwood TS-450S/950S to AEA PK-232 port configured for HF.

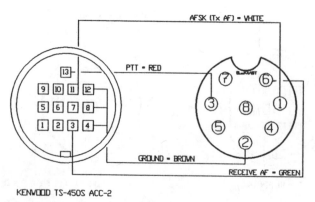

KENWOOD TS-450S ACCESSORY PORT "2"   KAM HF PORT.

T61. Kenwood TS-450S ACC-2 to KAM HF port.

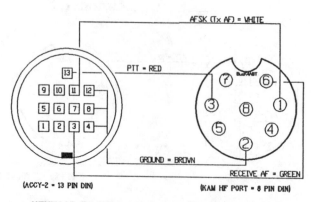

KENWOOD TS-690S ACC-2   KANTRONICS HF PORT.

T62. Kenwood TS-690S ACC-2 to Kantronics KAM.

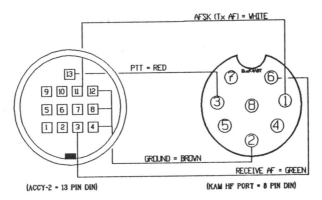

T63. Kenwood TS-711A ACC-2 to KAM HF port.

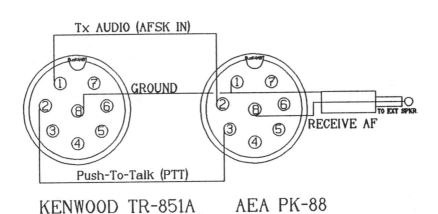

T64. Kenwood TR-851A to AEA PK-88.

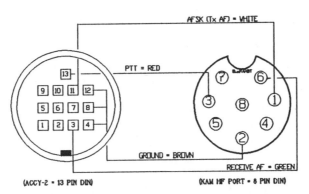

T65. Kenwood TS-950S ACC-2 to Kantronics KAM HF port.

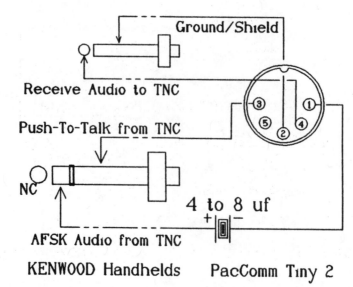

KENWOOD Handhelds   PacComm Tiny 2

T66. Kenwood handhelds (TH-25, 26, 27, 45, 55, 75, 205, 215, 225, 315, 415) to PacComm Tiny 2.

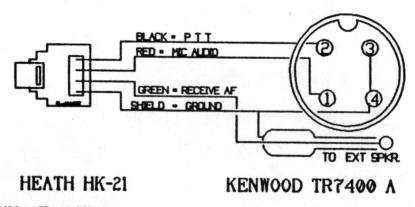

HEATH HK-21   KENWOOD TR7400 A

T67. Kenwood TR-7400 to Heath HK-21.

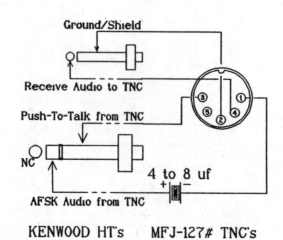

KENWOOD HT's   MFJ-127# TNC's

T68. Kenwood handhelds (TH-25, 26, 27, 45, 55, 75, 205, 215, 225, 315, 415) to MFJ-127# Series.

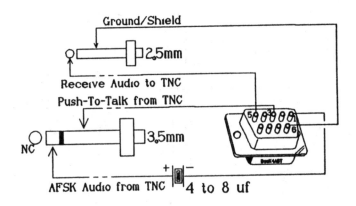

KENWOOD HT's   Kantronics KAM & KPC's

T69. Kenwood handhelds (TH-25, 26, 27, 45, 55, 75, 205, 215, 225, 315, 415) to Kantronics KAM KPC's.

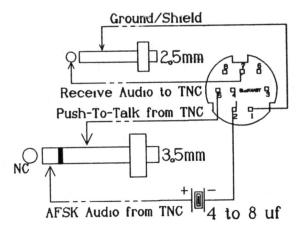

KENWOOD HT's   PacComm HandiPacket

T70. Kenwood handhelds (TH-25, 26, 27, 45, 55, 75, 205, 215, 225, 315, 415) to PacComm HandiPacket.

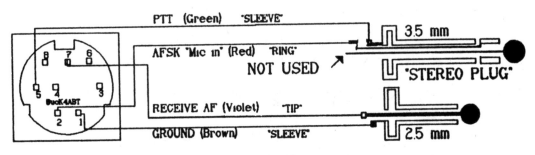

T71. Alternate method to interface Kenwood TH-25 to HandiPacket.

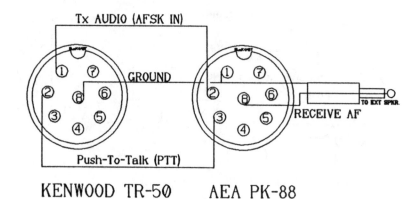

KENWOOD TR-50   AEA PK-88

T72. Kenwood TR-50 to AEA PK-88.

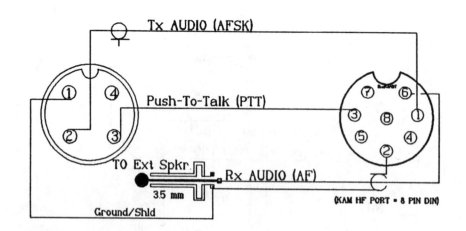

YAESU FT-101Z/ZD   KANTRONICS KAM HF PORT

T73. Yaesu FT-101Z/ZD to Kantronics KAM (HF) Port.

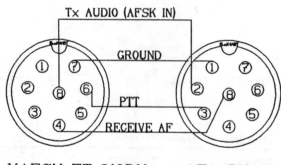

YAESU FT-212RH   AEA PK-88

T74. Yaesu FT-212RH to AEA PK-88.

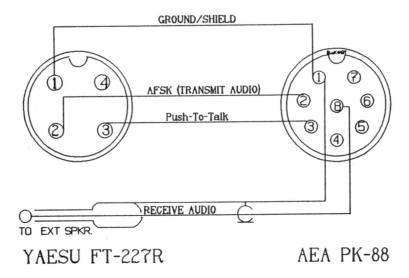

T75. Yaesu FT-227R to AEA PK-88.

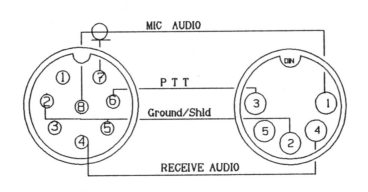

T76. Yaesu FT-2700 and 4700 to DRSI DPK-2.

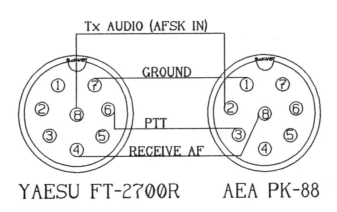

T77. Yaesu FT-2700R to AEA PK-88.

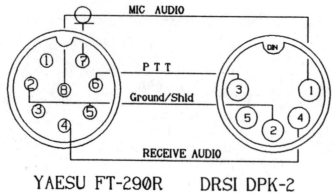

YAESU FT-290R   DRSI DPK-2

T78. Yaesu FT-290R to DRSI DPK-2.

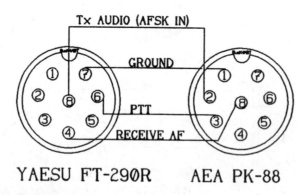

YAESU FT-290R   AEA PK-88

T79. Yaesu FT-290R to AEA PK-88.

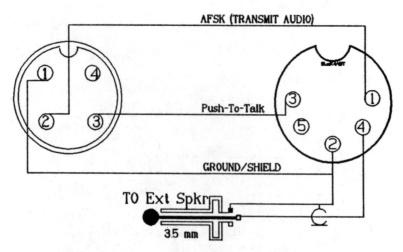

YAESU FT-301D   MFJ-1278 SET TO HF MODE.

T80. Yaesu FT-301D to MFJ-1278 (HF).

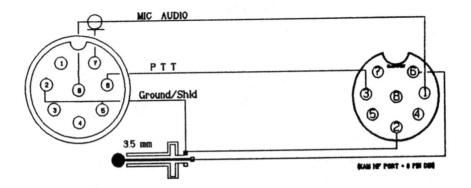

YAESU FT-747   KANTRONICS KAM HF PORT

T81. Yaesu FT-747 to Kantronics KAM (HF).

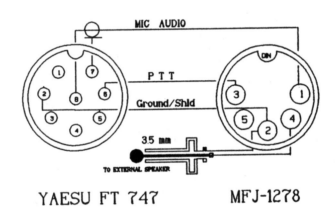

YAESU FT 747   MFJ-1278

T82. Yaesu FT-747 and 757 to MFJ-1278/Turbo.

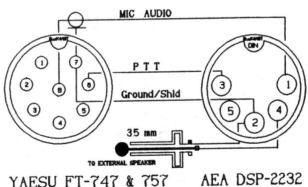

YAESU FT-747 & 757   AEA DSP-2232

T83. Yaesu FT-747 and 757 to AEA DSP-2232.

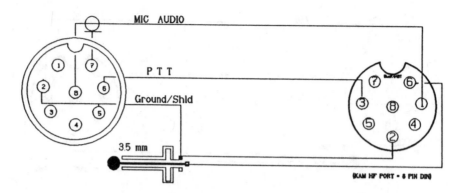

YAESU FT-757 GTX-II   KANTRONICS KAM HF PORT

T84. Yaesu FT-757 GTX II to Kantronics KAM (HF) port.

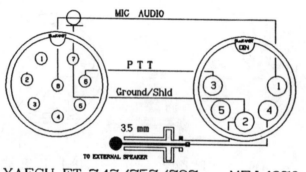

(Note: Configure MFJ-1278 for HF parameters.)

YAESU FT-747/757/767   MFJ-1278

T85. Yaesu FT-7#7 Series to MFJ-127# series.

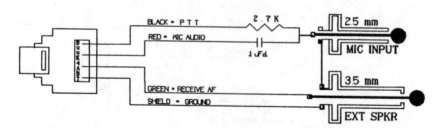

Heath HK-21 "Pocket-Packet"   YAESU FTs 209/709/727

T86. Yaesu FT-209/709/727 to Heath HK-21 ''Pocket-Packet.''

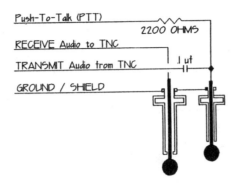

T87. TNC interface connections for the Yaesu HT models FT-23, 33, 73, 109, 209, 709, 727, 411, 811, 911, 470.

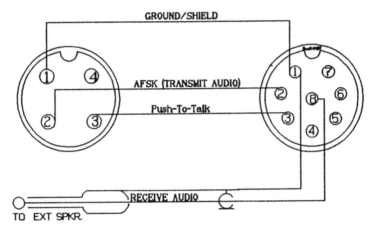

YAESU FT-101E   AEA PK-88

T88. Yaesu FT-101E to AEA PK-88.

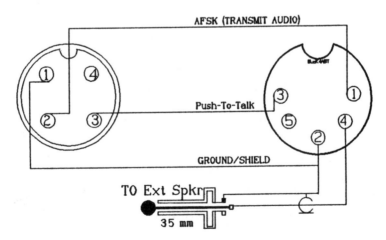

YAESU FT-101D/ZD   MFJ-1278 (Port set for HF).

T89. Yaesu FT-101D/ZD to MFJ-1278 (HF).

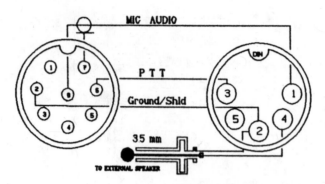

YAESU FT-221A, 321, & 421   PacComm Micro/Tiny 2

T90. Yaesu FT-221A and 421 to PacComm Micro/Tiny 2.

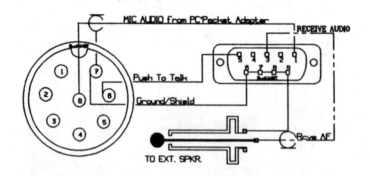

YAESU FT 747 HF TRANSCEIVER   DRSI PC*Packet Adapter

T91. Yaesu FT-747 to DRSI PC*Packet Adapter.

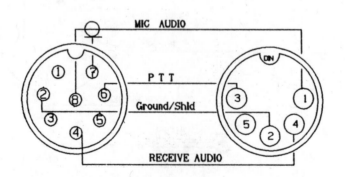

YAESU FT-211, 311, 411, & 2311   DRSI DPK-2

T92. Yaesu FT-211, 311, 411, and 2311 to DRSI DPK-2.

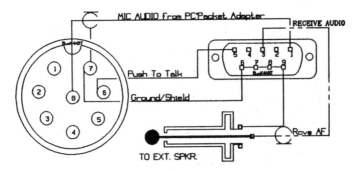

YAESU FT-757 GTX-II   DRSI PC*Packet Adapter

T93. Yaesu FT-757 GTX-II to DRSI PC*Packet Adapter.

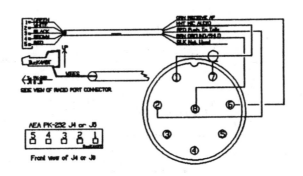

AEA PAKRATT 232 port 1 or 2.   Radio Shack HTX-100

T94. Radio Shack HTX-100 to AEA PK-232.

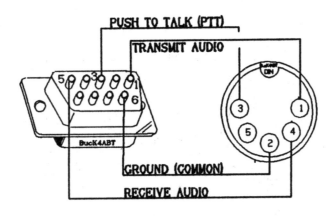

Kantronics D4-10 UHF transceiver   DRSI DPK-2

T95. Kantronics D4-10 UHF Transceiver to DRSI DPK-2.

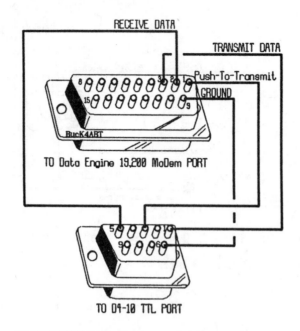

RECEIVE DATA

TRANSMIT DATA

Push-To-Transmit

GROUND

BucK4ABT

TO Data Engine 19,200 MoDem PORT

TO D4-10 TTL PORT

KANTRONICS Data Engine   D4-10 interface cable.

T96. Kantronics Data Engine to D4-10 Interface Cable.

FDK MULTI 2700

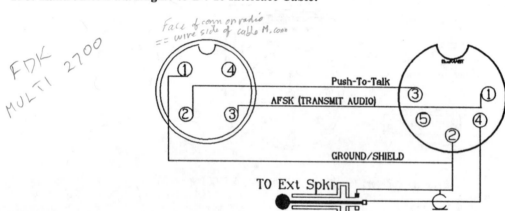

Face of conn on radio
== wire side of cable M, conn

Push-To-Talk

AFSK (TRANSMIT AUDIO)

GROUND/SHIELD

TO Ext Spkr

3.5 mm

KLM MULTI-2700   MFJ-1270B, 1274, & 1278/TURBO

T97. KLM Multi-2700 to MFJ-1270B, 1274, and 1278/Turbo.

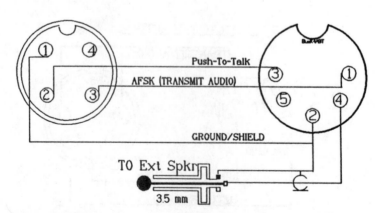

Push-To-Talk

AFSK (TRANSMIT AUDIO)

GROUND/SHIELD

TO Ext Spkr

3.5 mm

KLM "MULTI" 2700   PacComm TINY-2 & MicroPower 2

T98. KLM Multi-2700 to PacComm MicroPower and Tiny 2.

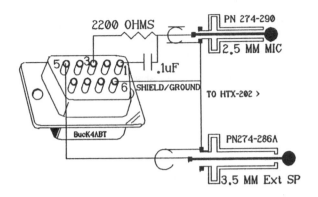

KPC-3   RADIO SHACK HTX-202 HT

T99. Radio Shack HTX-202 Handheld to Kantronics KPC-3.

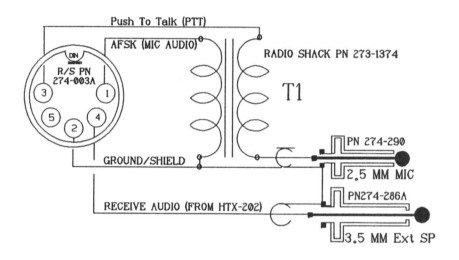

Note: The drawings shown
here are the same except that
I've drawn each in a different
manner to reduce confusion
when connecting to the isola-
tion transformer, T1.

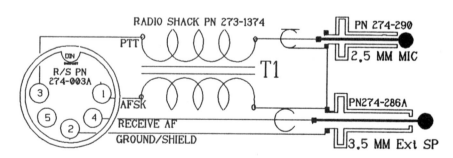

T100. Radio Shack HTX-202 transformer coupled to MFJ TNCs.

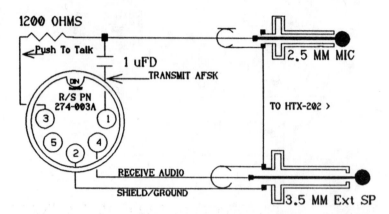

1200 OHMS

Push To Talk

1 uFD

TRANSMIT AFSK

R/S PN
274-003A

RECEIVE AUDIO

SHIELD/GROUND

2.5 MM MIC

TO HTX-202 >

3.5 MM Ext SP

TNC 2 or CLONE   Radio Shack HTX-202

T101. Radio Shack HTX-202 to TAPR TNC-2.

Computer Communications (Serial) COMPorts to TNCs

C1   DB-9 COMPort to DB-25 Serial Cable Adapter
C2   Most Commonly Used Signals of the RS-232 Standard
C3   Pin Definitions of the 25 Signals Related to RS-232
C4   MAC/SE to AEA PK-232 and PK-88
C5   Radio Shack Color Computers to AEA PK-232, 88, 87, MBX
C6   PC Compatible Serial (DB-25) to AEA PK-232, 88, 87
C7   PC Compatible Serial (DB-9) to AEA PK-232, 87, 88
C8   Atari Joystick Port to DRSI DPK-2
C9   Commodore C64 to DRSI DPK-2
C10   Radio Shack Color Computers to DRSI DPK-2
C11   IBM PC or Compatible (DB-25) to DRSI DPK-2
C12   PC Compatible (DB-25) to DRSI PC*Packet Adapter
C13   Early Apple to Kantronics KPC-3 or KAM
C14   Macintosh /SE to Kantronics KAM
C15   IBM/PC DB-9 or DB-25 to Kantronics Data Engine
C16   COMPAQ Laptop to Kantronics KPC-3 Pocket Communicator
C17   Radio Shack CoCo I, II, or III to Kantronics KPC-4
C18   PC or Compatible (DB-25) to Kantronics KPC-4
C19   Radio Shack CoCo's to Kantronics KAM, KPC-2, and KPC-3
C20   PC or Compatible to Kantronics KPC-2400
C21   PC (DB-9) COMPort to Kantronics KAM, KPC-2, and KPC-3
C22   Macintosh /SE to MFJ-1278
C23   Radio Shack CoCo's to MFJ-1270, 1274, and 1278
C24   PC or Compatible (DB-25) to MFJ TNCs
C25   Hewlett-Packard 95LX to HandiPacket (PacComm)
C26   Radio Shack Models 100 and 102 to HandiPacket
C27   Radio Shack Color Computers to PacComm MicroPower
C28   Hardware Handshake Interface, PC to PacComm Tiny 2
C29   Early Apple to TAPR TNC-2
C30   Commodore C-64 or Vic-20 to TAPR TNC-2 TTL Port

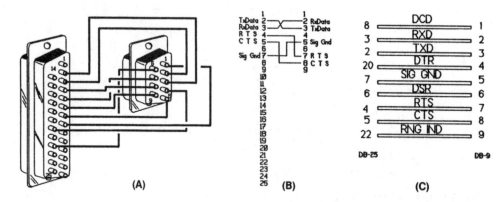

(A)   (B)   (C)

C1. DB-9 COMPort to DB-25 Serial Cable Adapter. Many late-model computers use the DB-9 connector at the serial port. At (B) the five most used signals are illustrated. At (C) all nine signals are used in the DB-9 to DB-25 adapter.

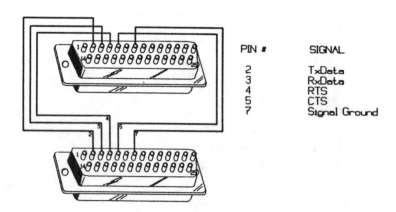

C2. DB-25 to DB-25, DTE to DCE, most commonly used RS-232 signals.

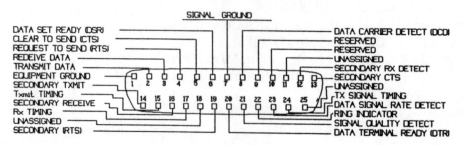

C3. The EIA RS-232 signals defined to the DB-25 connector employed on many personal computers.

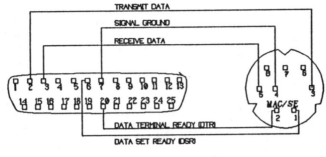

TRANSMIT DATA
SIGNAL GROUND
RECEIVE DATA
DATA TERMINAL READY (DTR)
DATA SET READY (DSR)

AEA PK-232 & PK-88   MAC/SE

C4. MAC/SE to AEA PK-232 and PK-88.

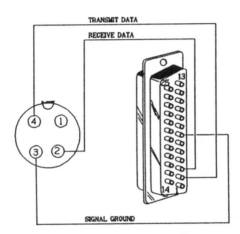

TRANSMIT DATA
RECEIVE DATA
SIGNAL GROUND

Radio Shack Color Computer I, II, & III

AEA PK-232/MBX, PK-87, & PK-88.

C5. Radio Shack color computers to AEA PK-232, 88, 87, and MBX.

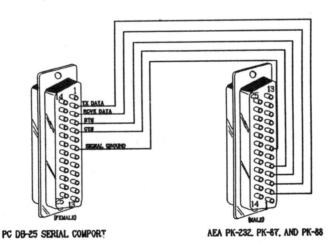

TX DATA
RCVE DATA
RTS
CTS
SIGNAL GROUND

PC DB-25 SERIAL COMPORT   AEA PK-232, PK-87, AND PK-88
(FEMALE)   (MALE)

C6. PC compatible serial (DB-25) to AEA PK-232, 88, and 87.

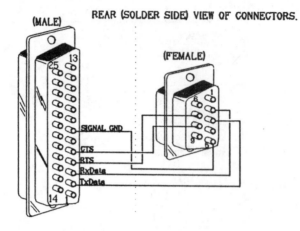

AEA PK-232, PK-87, & PK-88   PC OR COMPATIBLE SERIAL COMPORT

C7. PC compatible serial (DB-9) to AEA PK-232, 87, and 88.

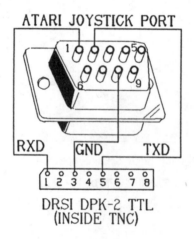

C8. Atari Joystick port to DRSI DPK-2.

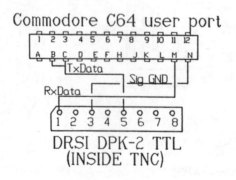

C9. Commodore C64 to DRSI DPK-2.

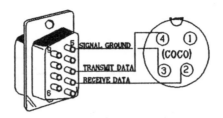

Radio Shack Color Computers I, II, & III
DRSI PC*Packet-Adapter & DPK-2

C10. Radio Shack color computers to DRSI DPK-2.

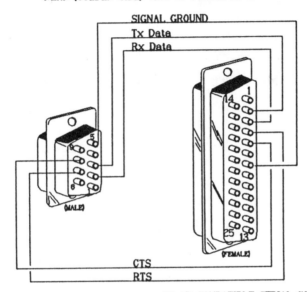

REAR (SOLDER SIDE) VIEW OF CONNECTORS.

SIGNAL GROUND
Tx Data
Rx Data

(MALE)

CTS
RTS

DRSI DPK-2   PC OR COMPATIBLE SERIAL COMPORT

(FEMALE)

C11. IBM PC or compatible (DB-25) to DRSI DPK-2.

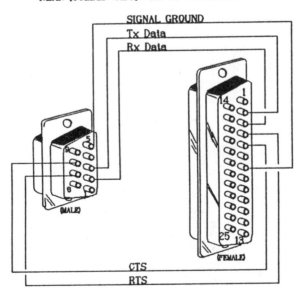

REAR (SOLDER SIDE) VIEW OF CONNECTORS.

SIGNAL GROUND
Tx Data
Rx Data

(MALE)

CTS
RTS

DRSI PC*Packet-Adapter   PC OR COMPATIBLE SERIAL COMPORT

(FEMALE)

C12. PC compatible (DB-25) to DRSI PC*Packet Adapter.

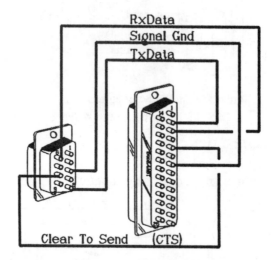

C13. Early Apple to Kantronics KPC-3 or KAM.

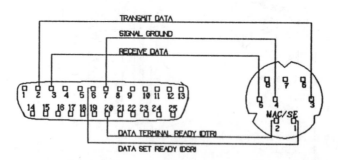

KANTRONICS KAM   MACINTOSH / SE

C14. Macintosh /SE to Kantronics KAM.

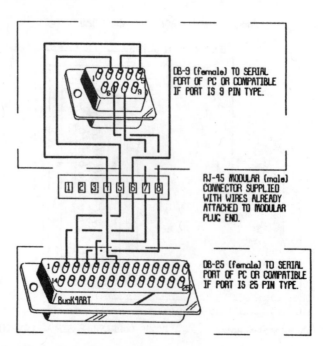

(Note: Configuration is shown for both types of computer serial ports—DB-25 and DB-9.)

C15. IBM/PC DB-9 or DB-25 to Kantronics Data Engine.

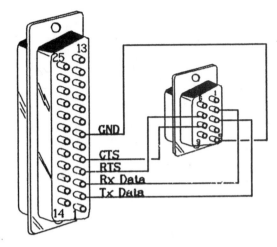

ComPaq LapTop   Kantronics KPC-3 Packet Communicator

C16. COMPAQ laptop to Kantronics KPC-3 Pocket Communicator.

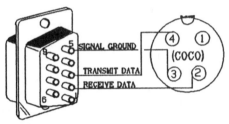

Radio Shack Color Computers I, II, & III   KANTRONICS KPC-4 (Dual-Port).

C17. Radio Shack CoCo I, II, or III to Kantronics KPC-4.

REAR (SOLDER SIDE) VIEW OF CONNECTORS.

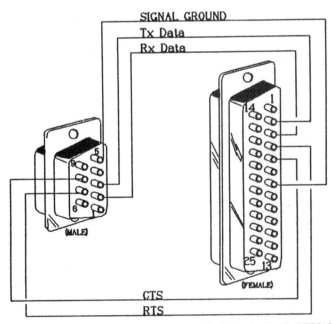

KANTRONICS KPC-4   PC OR COMPATIBLE SERIAL COMPORT

C18. PC or compatible (DB-25) to Kantronics KPC-4.

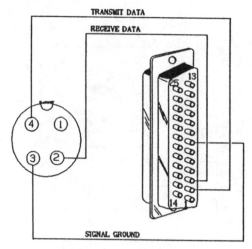

Radio Shack Color Computer I, II, & III   KAM, KPC-2, KPC-3, & KPC-2400

C19. Radio Shack CoCo's to Kantronics KAM, KPC-2, and KPC-3.

REAR (SOLDER SIDE) VIEW OF CONNECTORS.

PC DB-25 SERIAL COMPORT   KANTRONICS KAM, KPC-2, 3, & 2400's

C20. PC or Compatible to Kantronics KPC-2400.

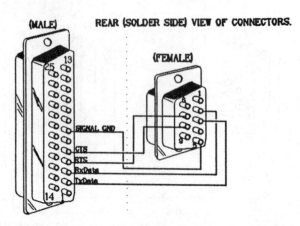

KANTRONICS KAM, KPC-2, & KPC-3   PC OR COMPATIBLE SERIAL COMPORT

C21. PC (DB-9) COMPort to Kantronics KAM, KPC-2, and KPC-3.

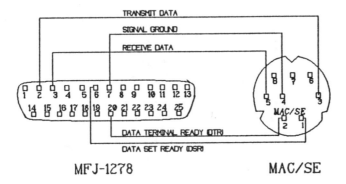

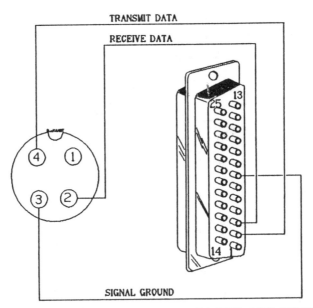

C22. Macintosh /SE to MFJ-1278.

Radio Shack Color Computer I, II, & III   MFJ-1278, MFJ-1270B, & MFJ-1274

C23. Radio Shack CoCo's to MFJ-1270, 1274, and 1278.

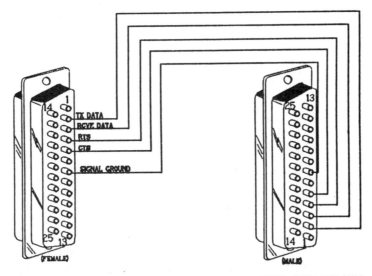

PC DB-25 SERIAL COMPORT   MFJ-1278/TURBO, 1270B, AND 1274

C24. PC or compatible (DB-25) to MFJ TNCs.

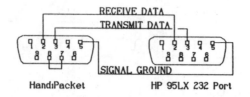

Note: The jumper at pins 7 to 8 must be added when using software handshaking with the HandiPacket.

HandiPacket   HP 95LX 232 Port

C25. Hewlett-Packard 95LX palm-size PC to HandiPacket (PacComm). A minor change is required when using the HP-95LX with the PacComm HandiPacket.

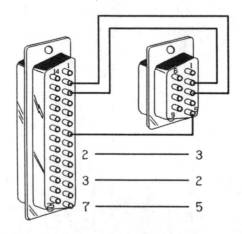

RADIO SHACK MODEL 100/102   PacComm MicroPacket

C26. Radio Shack Models 100 and 102 to HandiPacket. Set Model 100 or 102 "TELCOM" parameters to 57E1E or a terminal baud rate of 1200, with 7 data/word bits, EVEN parity, 1 stop bit, and ENABLE software handshaking. In most cases this should be compatible with the default parameters of the TNC.

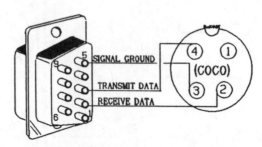

Radio Shack Color Computers I, II, & III
TO PacComm TINY-2, & Micro Power 2

C27. Radio Shack Color Computers to PacComm MicroPower.

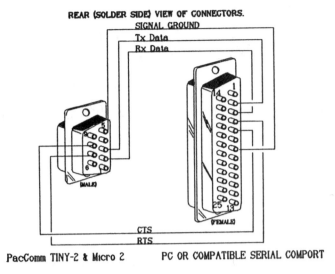

REAR (SOLDER SIDE) VIEW OF CONNECTORS.

SIGNAL GROUND
Tx Data
Rx Data

(MALE)

(FEMALE)

CTS
RTS

PacComm TINY-2 & Micro 2   PC OR COMPATIBLE SERIAL COMPORT

C28. Hardware handshake interface, PC to PacComm Tiny 2.

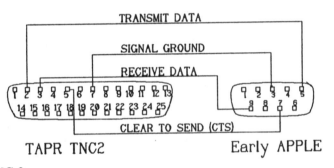

TRANSMIT DATA

SIGNAL GROUND

RECEIVE DATA

CLEAR TO SEND (CTS)

TAPR TNC2   Early APPLE

C29. Early Apple to TAPR TNC-2.

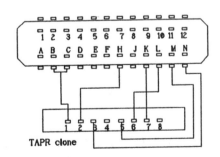

TAPR clone

C30. Commodore C-64 or Vic-20 to TAPR TNC-2 TTL port.

NOTES

Special Applications

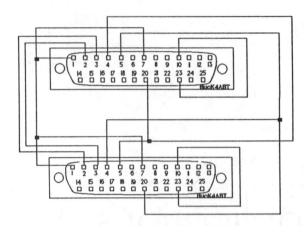

S1. This drawing illustrates how two nodes can be interfaced to annex LAN nodes to a "neighbor" node on the back-bone frequency.

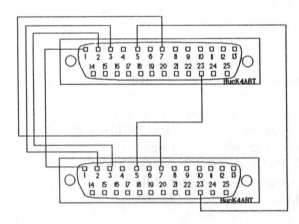

S2. Two ROSE switch interface LAN to backbone "gateway."

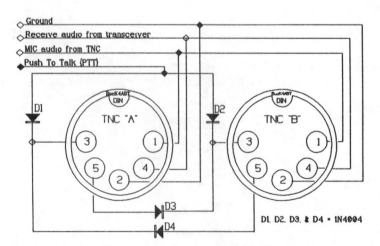

S3. Two TNCs to one radio interface "radio port." Use this special radio port cable when adding a 2400 bps node to an existing 1200 bps node. This configuration will not allow both TNCs to activate the PTT at the same time. It is also useful when adding the CONF node to the network node.

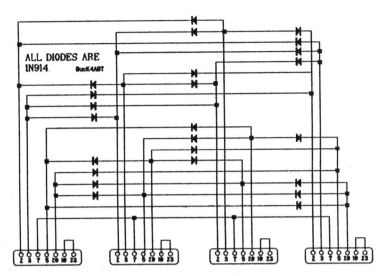

S4. Four-port LAN, trunk, and backbone diode matrix (TheNet). When more than two nodes are interfaced to link LANs, trunks, and backbones, it becomes necessary to construct the diode matrix shown here. (ROSE matrix is not the same.)

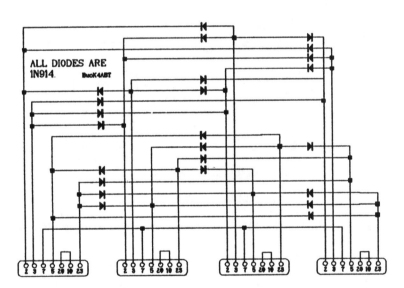

S5. Diode matrix used to interface up to four ROSE switches in a gateway configuration. Switches may be used to link to other frequencies, baudrates, or trunks. This matrix may be used to interface LANs into backbones.

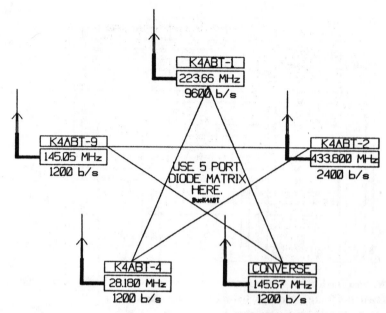

S6. The "star" configuration is a cluster complex that allows easy access to nodes, backbones, trunks, gateways, and CONFERENCE/CONVERSE nodes. Notice how the "star" allows connections from one frequency to another.

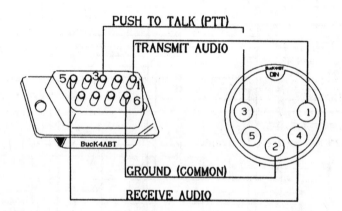

S7. Kantronics D4-10 UHF DataRadio to MFJ-1270B equipped with 9600 baud modem and ROSE firmware (EPROM) installed.

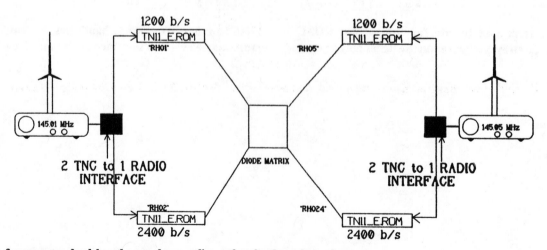

S8. Two-frequency, dual-baud complex configured as both node and gateway.

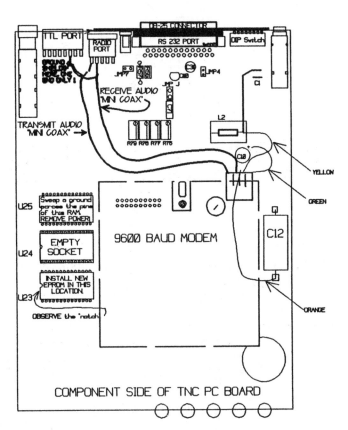

COMPONENT SIDE OF TNC PC BOARD

S9. TAPR TNC-2 clone showing installation and wiring of the 9600 baud modem. Install new EPROM (TheNet, Net/Rom, or ROSE) in IC socket U23. Be sure power is off and power cable is disconnected when performing the memory erase at IC socket U25. IC socket U24 remains empty. **When installing the EPROM at U23, make sure it is installed with the notch in the correct position. To install the EPROM backwards will destroy the EPROM when power is applied.**

S10. GE MVP Transceiver input-output connections. Notice that "receive" audio is present at pin 4 of the connector.

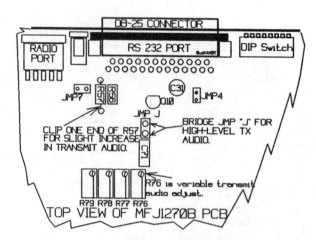

S11. Audio level mods, MFJ-1270B and MFJ-1274 TNCs to MICOR and GE MASTER II. It may be necessary to make one or more minor mods to the TNC AFSK output.

When a jumper is placed across "JMP J" the transmit audio output from the TNC may be more than 2 volts.

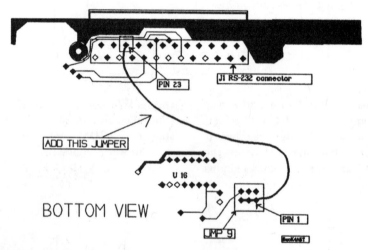

S12. TNC-2 mod to enable handshaking (node and ROSE). This minor modification enables flow control when two or more ROSE switches are connected back-to-

back or when connected to a diode matrix to form a multi-port gateway.

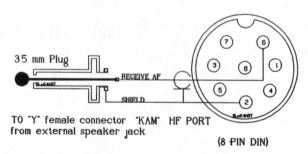

S13. A simple modification that enables the KAM's tuning indicator to be used at 1200 bps on HF. Use the VHF port and connect to the HF rig. Set the MARK and SPACE

commands as follows: MARK 1200, SPACE 2200. Above 28 MHz 1200 baud is legal.

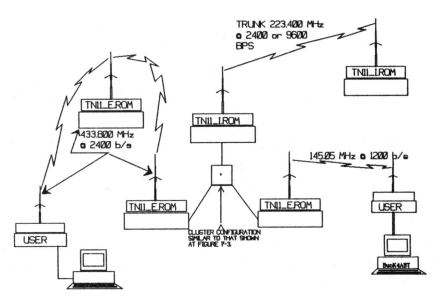

TRUNK 223.400 MHz
@ 2400 or 9600
BPS

TNI1_I.ROM

TNI1_E.ROM

433.800 MHz
@ 2400 b/s

TNI1_I.ROM

TNI1_E.ROM

TNI1_E.ROM

145.05 MHz @ 1200 b/s

USER

CLUSTER CONFIGURATION
SIMILAR TO THAT SHOWN
AT FIGURE 7-3.

USER

BuoK4ABT

S14. Limited access backbone with TheNet 2.10. User access limited by the use of T/N"I" firmware at critical nodes. Access is permitted at the T/N"E" type nodes. "LAAP" networks are becoming more common as the number of users increase.

(Note: At the SCOTTS-BORO switch the transition from 430.55 MHz to 223.580 MHz is similar to the way a LAN switch is interfaced into the backbone trunk.

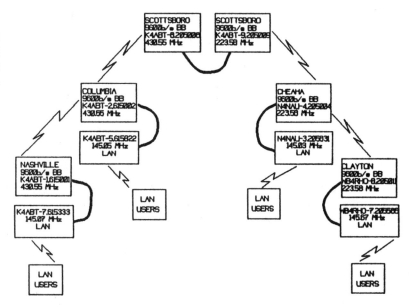

SCOTTSBORO
9600b/s BB
K4ABT-8.205006
430.55 MHz

SCOTTSBORO
9600b/s BB
K4ABT-9.205009
223.58 MHz

COLUMBIA
9600b/s BB
K4ABT-2.615002
430.55 MHz

CHEAHA
9600b/s BB
N4NAU-4.205004
223.58 MHz

K4ABT-5.615822
145.05 MHz
LAN

N4NAU-3.205831
145.03 MHz
LAN

NASHVILLE
9600b/s BB
K4ABT-1.615001
430.55 MHz

CLAYTON
9600b/s BB
NB4RHO-8.205011
223.58 MHz

K4ABT-7.615333
145.07 MHz
LAN

LAN
USERS

NB4RHO-7.205566
145.67 MHz
LAN

LAN
USERS

LAN
USERS

LAN
USERS

S15. Path and gateway routing example. Solid, curved lines indicate RS-232 interface between the LAN switch and the backbone. "Electric bolt" symbols indicate RF path between users and the LAN, and the backbone (BB) limited access switches.

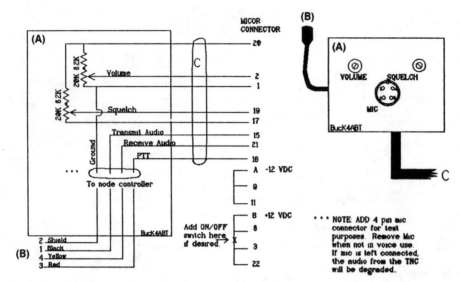

(Note: If TNC transmit audio is too low, clip one end of R-57 inside TNC-2 clone, or place jumper across JMP-J near Q10. The pin numbers at cable "C" represent the pins in the connector at the front of the MICOR and directly behind the handle.)

S16. Motorola MICOR interface for conversion to node/ROSE.

Index

NOTES

NOTES

NOTES

NOTES

NOTES